Edgar Mayhew Bacon, Eugene Murray-Aaron

**The New Jamaica**

Describing the Island, Explaining its Conditions of Life and Growth and Discussing its

Mercantile Relations and Potential Importance

Edgar Mayhew Bacon, Eugene Murray-Aaron

**The New Jamaica**
*Describing the Island, Explaining its Conditions of Life and Growth and Discussing its Mercantile Relations and Potential Importance*

ISBN/EAN: 9783337330439

Printed in Europe, USA, Canada, Australia, Japan

Cover: Foto ©berggeist007 / pixelio.de

More available books at **www.hansebooks.com**

*DESCRIBING THE ISLAND, EXPLAINING ITS CONDITIONS
OF LIFE AND GROWTH AND DISCUSSING ITS MER-
CANTILE RELATIONS AND POTENTIAL IM-
PORTANCE: ADDING SOMEWHAT IN
RELATION TO THOSE MATTERS
WHICH DIRECTLY INTEREST
THE TOURIST AND THE
HEALTH SEEKER.*

BY

# EDGAR MAYHEW BACON

AND

# EUGENE MURRAY AARON, PH.D.

ILLUSTRATED BY THE AUTHORS AFTER ORIGINAL SKETCHES AND FROM PHOTOGRAPHS
TAKEN BY DR. JAS. JOHNSTONE AND OTHERS.

NEW YORK
WALBRIDGE & CO.

———

KINGSTON
ASTON W. GARDNER & CO.

MDCCCXC

# CONTENTS.

# LIST OF ILLUSTRATIONS.

# PREFACE.

IN offering this book to the public the authors wish to acknowledge the many courtesies they have received from Jamaicans, and the efficient aid which has been given them in the labor of gathering material for it.

If the accomplished work falls short of the expectations of those who have so kindly encouraged its production, we can but say that from a mass of material we have tried to select that which will be of most value and interest to the reader ; leaving much unappropriated, not because the treasure is small, but because our vehicle is inadequate to carry it all.

We have tried especially to give a fair, comprehensive, though condensed, account of Jamaica as it is; its present life, its means of communication and travel, its growing commercial outlook and its intrinsic wealth.

As its history has been the subject of a cumulative literature, treating of the events of nearly four hundred years of unusual interest, it will be seen that any attempt to condense such a mass of material to the limit of two or three chapters must be prefaced with apology and received with indulgence. But as a knowledge of the more prominent facts and conditions of the older time are necessary to a clear understanding of the present life and promise of a new Jamaica, we can but assure our readers that this epitome has been as honestly and fairly prepared as possible.

Among the authorities consulted, the most prominent are Bryan Edwards' "History of the West Indies;" Chas. Leslie's "New Account of Jamaica;" John Esquemeling's "History of the Buccaneers;" Bridge's "Annals of Jamaica;" Washington Irving's "Life of Columbus;" Gardner's "History of Jamaica;" Montgomery Martin's "History of the British Colonies;" Rector of Port Royal's "Narrative;" several recent essays and magazine papers, and the "Hand-book of Jamaica." Besides these, we had access to the records and reports of the government printing office, kindly placed at our disposal; and various pamphlets, clippings and other material, furnished by private parties, whose names we would be glad to publish with our acknowledgment, did courtesy permit.

# INTRODUCTION.

THE approach to Jamaica, whether made by day or night, cannot fail to interest and charm the traveller. He must indeed be a blasé wanderer who can view the unfolding beauty of its mountains, the arboreal display of its coasts and the brighter green of its cultivations with indifference. Whether he first lands in the central regions of the north side, following the example of the Genoese, and enjoys the garden like beauty of St. Ann's ; or falls among the bananas of Port Antonio ; or seeks the south side and all its present and historic meaning, he is sure to find that fulfilment keeps pace with expectation.

Beyond the light-houses of Morant, and the foot-hills of St. Thomas ye East, rise the cloudy blue summits of the Eastern Mountains, the highest of them 7,350 feet above the sea level. There are the cocoa-nuts in long ranks upon the palisadoes, and among a number of islets and capes the vessel is piloted past the sheltering forts, into a bay that has been enriched with a thousand memories.

Jamaica lies between 17° 42′ and 18° 31′ North Latitude. Its total length is 144 miles and its greatest width 49 miles. It is politically divided into three counties—Cornwall, Middlesex and Surrey, which are further subdivided into fourteen parishes.

There are over thirty large bays and harbors, the finest being Kingston, Port Antonio, St. Ann's Bay, Montego Bay, Savana la Mar, Lucea and Port Morant.

The present government of the island, created by an order in council by her Majesty Queen Victoria, dated May 19, 1884, consists of a governor, appointed by the crown, and a council, part chosen by the people and part serving by virtue of official position. There is, first, the Privy Council, consisting of the Colonial Secretary, the Attorney General, the Director of Public Works and two others.

The Legislative Council, of which the governor is president, consists of the senior military officer in command of troops on the island, the Colonial Secretary, Director of Public Works, Attorney General, Medical Superintendent, Inspector of Schools, Collector General and eight elected members.

The civil establishment includes all officers and clerks in the department, enumerated above. The Judicial department consists of the Supreme Court, Administrator General, Island Record office, Vice Admiralty Court, Kingston Circuit Court and resident magistrates court.

Between 1866 to 1885 the governor annually appointed municipal and road boards as well as church wardens. But since the latter date a single parochial board for each parish has been instituted whose members, thirteen to eighteen in number, are elected for this office, with the exception of the person representing the electoral district in the council, and the custos of the parish, who presides. These parochial boards manage all local affairs. The Church of England in Jamaica was disestablished in 1870.

The corporate name of the Board of Kingston is the " Mayor and Council of Kingston."

The present governor of Jamaica is Sir Henry Arthur Blake, K. C. M. G.

# HISTORY OF JAMAICA.

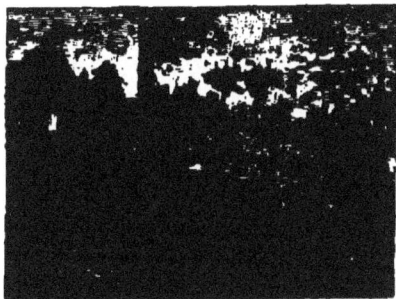

THE story of Jamaica's discovery by Columbus in 1494 is well known. It was on the great navigator's second voyage of discovery that he saw "the blue summits of a vast and lofty island at a great distance," which " began to rise like clouds above the horizon." Two days later, the hostility of the natives having been allayed, a landing was effected at a place which is called on the old maps Oracabessa. Its site is near the town of Port Maria, on a bay which the discoverer named Santa Gloria, in pious recognition of the beauty of the surrounding country.

Not until nine years later did the actual Spanish colonization of the island commence. When Columbus on a subsequent voyage, with caravels almost totally wrecked by the violence of an extraordinary tempest, was driven to seek a harbor, he put into the bay now known as Dry Harbor, which he called Puerto Bueno. That was on the 23d of June, 1503.

Excepting the gentle manliness of Columbus' own

character there is little that is pleasant to chronicle of the Spanish rule.

The earlier records of Jamaica are a chapter of more or less prosperous wickedness, seldom equalled in the world's story. The Spaniards, having done to death in various ways above sixty thousand of the aborigines, settled themselves down to the acquisition of wealth in their city of St. Jago, of which Columbus was created Duke. They had founded Melilla first, drifted thence to Savilla, abandoned Savilla to build Oristan, and this for St. Jago de la Vega—now Spanish Town.

The accounts of the Spanish occupancy having been written generally by Englishmen, are hardly matter to swear by. According to these narratives their lives were a compound of cruelty and indolence. If this was the case the English conquest in the time of Cromwell brought little material change to the land, except that the English drove a more flourishing trade with the outside world, and made their chief city at once a nest of vice and a centre of rude luxury.

In 1590 Sir Anthony Shirley, an Englishman, attacked the island and burned St. Jago, the capital, but did not choose to follow up his conquest. Upon the retirement of the English the Spaniards repaired Spanish Town, and were then unmolested by foreign foe till 1635. That year Colonel Jackson sailed with a small fleet to the Windward Islands and thence to Jamaica, where, with five hundred men, he attacked a garrison of two thousand Spaniards at Passage Fort, and after a hot fight, in which seven hundred Spaniards

are said to have been killed, utterly routed the settlers.

Having visited Spanish Town and extorted ransom,
Jackson followed Shirley's example and retired. But
a few years later Jamaica was again taken. "The
Crafty Mazerine,"—to quote Charles Leslie's honest
Royalist Chronicle,—" having engaged Cromwell to
join with France and turn his arms against Spain,
*politickly* contrived an expedition to the West Indies."
A fine fleet was fitted, aboard of which were " two
thousand old cavaliers and as many of Oliver's army."
The commanders were Colonel Venables and Admiral
Penn, the father of William Penn, who got one thousand three hundred more adventurers at Barbadoes and
the Windward Islands. With such an army, good
ships and able officers, they attacked St. Jago, after an
unsuccessful expedition against Hispaniola. In May,
1655, St. Jago capitulated to this force, its forts and
defences proving all inadequate against the munitions
of the invaders. But while parleying, and amusing the
English with fair speeches and presents, the Spaniards
contrived to remove much of their treasure from St.
Jago, and the same is supposed by treasure seekers and
other romantic people to be hid to this day in wells
and other safe places in the neighborhood.

After the English had gained the city they were
afraid of the foe, who still retained possession of the
country and greatly harassed them by sudden sorties
and skirmishes. At length, however, the conquest was
complete. The last Spanish governor fled to Cuba,
from a point on the north side of Jamaica, still known

as Runaway Bay. From this time British rule was permanently established.

When Admiral Penn and Colonel Venables returned to England they left in charge of the colony Colonel

D'Oyley, whose command included nearly three thousand men and twenty war vessels. D'Oyley was a brave and excellent leader. It was through him that the last remnant of the Spaniards were driven

from the island. But they left behind them a number of slaves, probably of mixed Carib and African blood, who, being fierce and warlike, took to the mountain fastnesses and became bandits, preying upon the fields and endangering the persons of the new settlers. D'Oyley succeeded in subduing them for a time, but he left a few individuals, who in later years grew to be powerful and greatly harassed the colony. Their part in Jamaican history has been sufficiently prominent to warrant this account of their origin. A remnant of them is still left, peacefully enjoying the privileges and immunities which they formerly wrested from the government. They are known as the Maroons.

Cromwell fitted a second squadron and sent Major Sedjwick to relieve Colonel D'Oyley. Before Sedjwick's arrival D'Oyley suppressed a mutiny among his men, shooting the ringleaders.

The new governor lived but a few days after his arrival, and the popular cavalier again resumed the direction of affairs.

Cromwell then appointed Colonel Brayne, of Scotland, with orders to colonize one thousand roundheads from Port Patrick to balance the royalists of D'Oyley's party. But Colonel Brayne followed Sedjwick, and for a third time D'Oyley ruled. He was a wise and energetic leader, governing with forethought and prudence. Having been twice supplanted by Cromwell, because he was a Royalist, he was finally removed by Charles II. upon his accession to the throne to make place for the royal favorite, Lord Windsor;

leaving so good a reputation, however, that he was long looked upon as the best of the governors.

. The new governor did little, but, to quote Chas. Leslie's venerable history, "In my Lord Windsor's government the Island was in a very flourishing condition, for by this time the buccaneers had begun their trade of pyrating *and made money plentiful.*" About this time, too, there were many wealthy men who came from other islands to settle in Jamaica. Among these was Sir Thos. Moddiford, afterwards governor. Sir Chas. Lyttleton followed Windsor, the latter being removed finally at the earnest protest of the Spaniards, who complained bitterly of the part he took in protecting the pirates. Under Lyttleton the first concessions were made to the Maroons, grants of land and magisterial power being given to Juan de Bolas, their leader. The governor also issued writs for the first general assembly held upon the island. Members were returned from twelve districts and met at Santiago de la Vega (now Spanish Town), where they indulged in great conviviality, if we may trust the older histories.

This first assembly was dissolved by Deputy Gov. Sir Edward Morgan. Following him came Moddiford, whose rule, says one of the chronicles, "brought the island to its greatest perfection." The population was then 17,298 inhabitants. Money was plenty, immigration increased and affairs were generally in a prosperous condition. Writs were issued for a new council, which proved to be rather combative in its temper than deliberative. One of its members murdered another at a state dinner.

While the assembly were quarrelling, the governor, on his own responsibility, was amusing himself by granting commissions and letters of marque to the pirates who already swarmed the Spanish Main. These were to annoy the fleets of Spain.

No chapter in the world's annals presents more appropriate material for modern melodrama than the lives of the buccaneers.

Bartholomew, a Portuguese, was the first buccaneer of note and achieved some brilliant successes, but was soon overshadowed by others. Brafiliano, a Dutchman, took some valuable prizes and greatly harassed the Spaniards. Lewis Scott was the first to land a force on Spanish territory and engage in terrestrial warfare, one of his acts being the sack of Campeche. Mansvelt took the Island of St. Catherines and wanted to hold it, under colonial protection, as a pirate rendezvous. He extorted a great ransom. The redoubtable John Davis carried fire and sword into Nicaragua and St. Augustine, retiring with immense booty. But the greatest of all the buccaneers was Henry Morgan. The son of a poor Welsh farmer, sold into servitude in Barbadoes and serving his term of slavery as a laborer, he impressed upon his time a romantic enthusiasm for his deeds and personality. Although greatly admired and copied by other privateers, Morgan is said by his biographers to have been unlike them, though in what the dissimilarity consisted we of a later day may be too dull to discover.

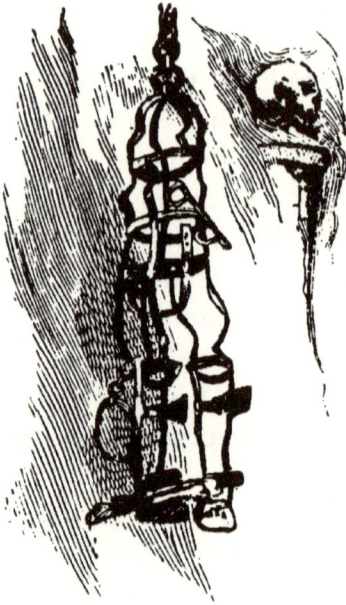

By his followers were committed cruelties unexampled; yet he is spoken of as being on a moral plane far above such men as Mansvelt, with whom, by the way, he sailed as vice-admiral in the latter's successful expedition against St. Catherines. Morgan, upon the death of Mansvelt, became the great pirate leader. He never sailed without a commission, however, and so over his colossal barbarities was thrown the cloak of authority, and expeditions for pillage and rapine were dignified as naval encounters and invasions.

In 1670, with an army of 1200 men and a numerous fleet, he attacked the town of Panama, then very rich; was victorious over the army that was sent against him and secured 175 mule loads of precious metal. Of this plunder his crew received only two hundred pieces of eight each, and mutinied, whereupon this intrepid leader stole away with treasure to the value of £25,000.

The immense wealth at this period brought into Port Royal; the thousands of freebooters whose money, bought with blood, was spent in crime; the cargoes of merchant fleets brought to its stalls and the ransom of provinces paid into its coffers, made this city enormously wealthy. Its state was barbaric but splendid; no form of vice was wanting, no indulgence too extravagant for its lawless population.

One of the curious contradictions of history occurred about here. Sir Thos. Moddiford was relieved, and sailed for England as a prisoner, to answer for the offence of exceeding his authority in commissioning Morgan. About the same time Morgan was knighted for his victory at Panama and was thereafter known as Sir Henry Morgan, the wealthy planter, the foe of the pirates and the friend to law and order.

Six years later Morgan, as Lieutenant-Governor, assumed control of Jamaica's affairs and was exceedingly popular.

Over a thousand Surinamese Dutchmen immigrated to the island in 1672. They were of industrious habits, and added to the colony's prosperity. A general awakening to industry resulted in the first shipment of sugar to England, the beginning of a trade which was for years the fruitful source of wealth to the colony, and which, a century later, brought Jamaica to the zenith of her prosperity.

The final crushing of the pirates and the unpopularity consequent upon the financial depression which followed, belonged to Lord Vaughn, who recalled the buccaneer's commissions and hung a great many of those marauders, thus effectually suppressing the dreadful business. It was at that time that the Royal African Company gained their charter which gave them every advantage upon the high seas, so that the

Jamaica slave trade was seriously interfered with, and the price of human flesh rose enormously.

In 1678 the Earl of Carlisle summoned a new assembly. Both he and his successors were perpetually in hot water, standing often between the colony and the mother country, on questions of financial policy principally.

When the Duke of Albemarle came, he established a claim to historic mention by bringing with him a great man, Sir Hans Sloane, the naturalist. The work of this extraordinary person, though accomplished before the discovery of our modern system of classification in Natural History, was of immense benefit to science, and stands to-day a monument and a landmark in a history of moral degradation, intellectual barrenness, political errors and mercantile obliquity.

The flight of James II. and the accession of William and Mary to the throne of England intensified for a time the political differences, which never were allowed to die. Certain acts, inimical, it was claimed, to the interests of Jamaica, were repealed and the constitution restored, which had been changed in Albemarle's time. To give the details of the perpetual wrangling which agitated Jamaica's rulers year after year would be neither interesting nor instructive.

The Earl of Inchequin, who took charge in 1690 varied the usual order of quarrel by sending the war ships *Severn* and *Guernsey* to retaliate upon the French, who had been annoying the sea-coast inhabitants of the island. These vessels took valuable prizes in Hispaniola. But Inchequin did not live to enjoy the prestige which such success usually brings.

We now come to one of the most memorable events in Jamaican annals.. On the 7th of June, 1692, a great earthquake shook the island and almost totally destroyed the Metropolis. Mountains were riven, earth and rock fell upon the valleys, burying the people, hamlets were engulfed, plantations obliterated and rivers turned into new channels.

The terrible retribution that overtook Port Royal in three or four brief minutes of time can be only compared in magnitude to the unexampled record of her debauchery. It was a disaster which in a moment transformed the richest spot on earth to the poorest. Even Lisbon's fate could not compare with the complete overthrow of the Jamaican capital. Leslie says: "At the Time when the Island was full of Gay Hopes, Wallowing in Riches and Abandoned to Wickedness, the most dreadful Calamity befel it that ever happened to a people, and which many look upon as a tremendous judgment of the Almighty. On the 7th of June, 1692, one of the most violent earthquakes happened that perhaps was ever felt. It began between 11 and 12 o'clock at noon, shook down and drowned nine-tenths of *Port Royal* in two minutes time; all the Wharves at *Port Royal* sunk at once. There were soon several Fathoms of Water where the Streets stood; and that one which suffered the least Damage was so overflowed that the Water swelled as high as the Upper Rooms of the Houses." Added to all the other horrors, the unburied dead which lay in heaps upon the land or floated in shoals in the harbor, became in a little while, under that tropic sun, horrible masses of putrefaction, generating a pestilence from which thousands of those who had survived the earthquake died.

The overthrow of Port Royal led to the establishment of the city of Kingston on the Liguanea Plain, upon property belonging to Colonel, afterward Sir William Beeston. The city was laid out by Colonel Christian Lilly, of the Royal Engineers.

Shortly after these events Beeston assumed the government (in 1693). It was then that the French were again peculiarly active and annoying. They had burned plantations in Jamaica, and taken away slaves to the value of £65,000. The colonial militia finally succeeded in defeating these invaders on the land, driving them back to their ships with loss ; but on the water the French were victorious, and the great English Admiral Benbow was defeated, dying from his wounds in Kingston shortly afterwards.

During several administrations the usual succession of legislative troubles engaged the attention of the governors. The Picaroons from Cuba created a diversion in the time of Sir Nicholas Lawes, by committing many depredations, and the embarrassment thus caused to agriculture was further augmented by a hurricane, which destroyed both lives and property. Yet the government could hardly leave its wrangling over the question of a permanent revenue bill long enough to take proper measures for the relief of the sufferers.

Then followed a ruler whose course of conduct, being in marked contrast to those who had preceded him, demands recognition. Major-General Robert Hunter, learning that he was about to receive the appointment to Jamaica, actually took pains to inform himself of the condition of the country and people to which he was going, and so effectually presented their case and cause to his Majesty's ministers as to win certain concessions for them. The Jamaica assembly, feeling that the country had a friend in the new governor, promptly passed the much discussed bill, granting a permanent revenue of £8000 per annum to the crown, receiving in return the confirmation of their laws, for which they had been fighting. Besides this, Hunter's salary was increased from £5000 to £6000 as a token of gratitude for his services.

Trouble with the Maroons, already referred to, now reached its height. Many skirmishes were fought, and the whole colony was in a state of insecurity and alarm. Indeed, the Maroon war lasted with greater or less intensity for nearly forty years, and it was not till Governor Trelawney's arrival that peace was concluded with these mountaineers. By grants of land and peculiar privileges, he succeeded in forming a treaty which was in all points as though concluded with a foreign power, instead of with a band of predatory savages in the act of rebellion.

In 1739, the war between England and Spain called out a volunteer force from Jamaica to assist against the South American ports. The expedition in which they engaged led to the surrender of the

Spanish American towns of Chagres and Porto Bello.

During Trelawney's administration in 1744, another earthquake shook Port Royal, and a great hurricane and tidal wave swept Savanna la Mar so that the place, people, houses and cattle were utterly destroyed.

Governor Knowles, in 1751, was burned in effigy for some differences with the House. In 1760 a slave insurrection broke out in the parish of St. Mary. Whole families of white planters were butchered by the insurgents, and it was only after a battle during which four hundred of them were killed that peace was restored. The ringleaders were shot or hung in chains, and many of the others transported.

In 1762, Governor Lyttleton brought news of another war between Spain and England. An expedition sent against Havannah was successful, and that city capitulated. Besides this victory, the capture of twelve ships of the line, and a fleet of merchantmen, swelled the amount of booty to £2,000,000, and made Jamaica rich once more.

In the time of Elletson, who succeeded Lyttleton, another negro outbreak occurred in Hanover and Westmoreland. It was stamped out, and thirty ringleaders were hanged. Soon after this the political world was agitated over the American war for independence, the recognition of the United States by France, and the consequent war between that country and Great Britain. Martial law was proclaimed in Jamaica, and the principal ports of the island were fortified. Nelson, who was then commander of Fort

Charles, volunteered in an expedition against Nicaragua, and nearly lost his life. Admiral Rodney, Jamaica's best loved hero, won a great victory over the French Admiral Du Casse, in April, 1782, thereby saving the island from a troublesome foe, and winning for himself the thanks of his sovereign, and his elevation to the peerage. Rodney's statue, by John Bacon, now occupies a prominent position in the public square at Spanish Town.

Following these troublous times, Jamaica was plagued with famine, and swept by hurricanes for the space of several years.

The year 1795 saw another formidable Maroon outbreak, at the termination of which six hundred of these troublesome neighbors were transported to Nova Scotia.

During the eighteenth century the population of the island had greatly increased, and as towards the close of the seventeenth, the great wealth brought by the buccaneers had given a dazzling though temporary and fictitious prosperity to Jamaica, so the closing decades of the eighteenth saw this wealth and luxury repeated upon the, apparently, more stable foundation of agriculture and commerce.

In spite of legislative brawls, and the dangers resulting from an isolated, almost defenceless condition, the "Gem of the Antilles" was enjoying her age of gold at the commencement of the present century.

During the eighteenth century the importation of human cattle from Africa reached 600,000 souls. The mortality among them must have been very great, for in spite of their natural tendency to increase, the close of the slave trade found barely half that number on the island. Bryan Edwards says: "It appears to me that the British slave trade had attained its highest pitch of prosperity a short time before the American War" (the War for Independence is referred to). The number of ships which sailed from England to the coast, engaged in the nefarious business of slave trading in 1771, was 196; and the total number taken to British colonies in that year (of which Jamaica took the lion's share) was 47,146. The treatment these poor creatures received at the hands of their masters was often brutal, and nearly always, to state it mildly, unsympathetic. This will be referred to further on as one of the potent causes of difficulty between the different classes of the population.

A mutiny among the troops occurred during William, Duke of Manchester's administration of the government, and troubles multiplied. Wars interfered with commerce, storms devastated the plantations, and the agitation over the slave question became more and more violent.

The bitter feeling of the planters against the Imperial Government resulted in a threat to unite with the United States. The excitement spread to the slaves. An outbreak and bloodshed was the result, and martial law was proclaimed. During the Earl of Musgrave's rule, the colony denied the right of the Imperial Government to legislate for Jamaica. A long controversy ensued, resulting in the passing of

the Emancipation Act, which provided that, " From ·
and after the 1st of August, 1834, all the slaves in the
colonial possessions of Great Britain should be forever
free, but subject to an intermediate state of six years'
apprenticeship for prædials and four years for do-
mestics." In 1838 and '40 the negroes of Jamaica,
through the exertions of the venerated Wilberforce
and others, became freedmen. In the early years of
one of the greatest reigns that England has known,
this attempt was made to right a great wrong. In the
fifty years that have intervened, the experiment has
been working, at first very slowly, because of dense
ignorance and great misunderstanding on both sides,
but latterly more rapidly toward its legitimate conclusion.

The history of Jamaica since the year 1840 is almost too recent to be fairly written. It is difficult to
get the proper proportion and perspective at such
short range. Unquestionably the immediate effect of
emancipation was disastrous, coming when it did,
upon Jamaican industry. Before that time the estates were greatly impoverished, and were beginning
to yield much less than they had done a few years
previously. In 1805, had been the largest production
of sugar, the estates aggregating that year 150,352
hogsheads. This was doubtless partly due to the
introduction of Bourbon cane in 1799, for at the beginning of this century the output of sugar rose from 95,-
858 to 110,000 hogsheads, and never fell below 100,000
again till Mr. Channing made his anti-slavery resolutions in 1823. Rum, too, in 1806 reached 58,780

puncheons. Fifty years later the production of sugar
had fallen to 41,656 hogsheads.

Emancipation found the planters in a pitiable condition financially. The majority were debtors to
English houses. The £5,853,975 sterling awarded as
compensation for the loss of their human property,
insufficient as the sum was, went for the most part
into the hands of their creditors. They were left
without resources, with overworked estates, antiquated
machinery, scarcity of labor and a poor market.

Lord Sligo, who arrived in 1835, found his part in
an impoverished country, a thankless one. He soon
gave place to Sir Chas. Metcalfe, who succeeded in
restoring peace between Jamaica and the mother
country. He retired in 1842.

During these years, further misfortunes visited the
planters. In slavery times the English Government
by heavy differential duty on foreign sugar, protected Jamaica. But the adoption of free trade
policy a few years after emancipation reduced the
price of sugar one-half to the English consumer, and
made the planter's profit correspondingly lighter at a
time when he could ill afford any diminution of in-
come. Abolition had cut down the labor supply.
Free trade had further diminished the chance for
profit in sugar growing. Estates were heavily mort-
gaged and many were abandoned.

To understand more fully the condition of the
island at this time, and its bearing upon subsequent
events, it will be necessary to inquire what were the
relations existing between the white man and the

negro before and after the latter ceased to be property. In the early days the slaves were undoubtedly
overworked and cruelly treated. In the years suc-
ceeding those of the two greatest crops, nine thousand were annually imported to repair losses. Since
emancipation the freedmen have multiplied by natural
increase.

Rum of Sc:............ill - ................

Each slave had a little patch of ground which he
was allowed to cultivate ; he was given two suits of
clothes per annum, and provided with medical attendance when ill. The average value of a slave was in
the neighborhood of £35. England paid £19 per
capita for them.

Doubtless there was much reason for the exceedingly bitter feeling with which the two classes were inclined to regard each other.

There appears to have been a downright fear on the part of the white Creole that the black men might assert their numerical superiority and take matters on the island into their own hands. That seems to be the reason for certain discriminating laws by which planters could eject negro tenants at a week's notice and destroy the fruits of their industry. There was a heavy stamp duty on the transfer of small parcels of land; an import duty on corn food, which was raised from 3*d.* to 3*s.* per barrel; a duty on shingles (while the staves and hoops used by the planter had the duty to which they were subject, reduced); a license required of those who sold at retail, while none was required of the wholesale dealer; a discriminating tax on sugar and coffee.

In addition to these legal bars and checks the planters refused to sell or lease small holdings, so that the negroes were compelled to wait for estates to go to the hammer.

These things were done, apparently, for no other reason than to prevent the exercise of that brute power which the negro undoubtedly possessed. Outnumbered in a country where the inhabitants were subject to every insular disadvantage, it is not strange that a violent use was made of such strength as the white man had to keep his black neighbor disarmed.

It must be borne in mind that the abolition act of May, 1833, followed the slave uprising of the previous year and the excitement that attended it; an agitation so great that certain dissenting ministers and missionaries were freely charged with incendiary conduct. In 1840 a plan for the introduction of coolie laborers was carried into effect, 20,000 of these Indian hands being thus added to the working population of the island. It was not only expected that the coolie would supplement the ordinary labor supply, but would, upon extraordinary occasions, stand between the planter and the inconvenience and loss which he experienced from the intermittent industry of the irresponsible freedman. The Indian was a check upon that spirit of independence which, however commendable in theory, has sometimes been a bane practically.

The introduction of the coolies, like the acts immediately following the abolition of African slavery, was simply an expedient; a bridge by which the governing class tried to cross that slough of despond by which Jamaican industries were encompassed. In spite of the honest opposition to which it was subjected it bids fair to prove itself an act of statesmanship, having resulted in the permanent accomplishment of several of the results sought for. As to the other acts to which we have referred, they did not even serve their immediate purpose. Indeed, the various taxes, checks and disabilities to which the negroes were then subjected hastened an outbreak which culminated in a veritable reign of terror.

In 1850, the island was cursed with its first infliction of Asiatic cholera, which nearly decimated the population and further depressed agriculture and commerce.

In 1865, while Mr. Edward John Eyre was governor of Jamaica, a storm which had been long gathering, burst upon the island. While those who participated in the events of that time are still, in many cases, engaged actively in the government and social affairs of the colony, it is yet too early to give more than a brief outline of the negro uprising in the East.

Mr. George William Gordon, born a slave and the son of his master, had become a man of mark in Jamaica, having acquired property and being actively engaged in politics. An elected member of the government, and belonging by virtue of his possessions to a class usually conservative, Mr. Gordon was called by many people an agitator—a negrophile.

At a time when meteorological causes had resulted in poor crops, and the American war made provisions high, so that the people were discontented by reason of actual want, Gordon presided at a meeting in Kingston at which speeches were made inciting, urging the people of African descent to "form themselves into societies, hold public meetings, and co-operate for the purpose of setting forth their grievances."

Whatever the purpose of Gordon and his party, it was soon lost sight of in the disastrous and unlooked-for result. The people to whom he had appealed, being very ignorant, knew nothing of arguments or appeals, or the niceties of legal redress. They were abundantly gifted with savage passions, and they were proficient in the use of the machete. There were certain individuals whom they greatly hated, and a class whose interests were all opposed to their own.

They would appeal to the machete. That seemed reasonable to them.

On the 11th of October the custos and vestry of St. Thomas in the East, met at the court-house at Morant, where they were attended by a protecting body of volunteers. An attack by the excited blacks resulted in the murder of nearly all of that vestry, the slaughter of all the officers and nearly all of the private men of the volunteer command, and the perpetuation of the most atrocious barbarity by the insurgents.

The fight at Morant Court-house was one of almost unexempled ferocity and horror. The pillage, arson and bloodshed which followed it filled the island with terror.

When Governor Eyre was informed of the outrage he took measures to put down the rebellion. White troops and volunteers, aided effectually by the Maroons, crushed the insurrection in a week. Martial law had at once been proclaimed and was in operation for a month.

Among the hundreds who were arrested was Mr. Gordon, who was summarily tried by court-martial and on the verdict of that insufficient tribunal, hanged. Those of his friends and enemies now living may settle upon the term by which his execution is to be named.

The report of the commission appointed by the crown to inquire into the cause of the outbreak, and the means used to suppress it, will give a clear enough idea of Governor Eyre's acts. The commission of inquiry, whose president was Sir Henry Knight

Storks, associated with whom were Mr. Russell Guer-
ney, the Recorder, of London, and Mr. J. B. Maule,
the Recorder of Leeds, reported as follows :

" (1) That the punishments inflicted during martial
law were excessive ; (2) that the punishment of
death was unnecessarily frequent ; (3) that the flog-
gings were reckless and at Bath positively barbarous ;
(4) that the burning of one thousand houses was
wanton and cruel." The commissioners also reported
that the " disturbances had their immediate origin in a
planned resistance to lawful authority," and that " a
principal object of the disturbers of order was the
obtaining of land free from the payment of rent."

# LATER DAYS.

THERE is every reason for refraining from presenting for the perusal of intelligent men the record of their own acts, except when they have become components of a rounded epoch, a completed chapter of history.

The years following the insurrection that ended with the execution of Gordon were marked by some radical changes, both in the form of government and the commercial life of the colony.

Under Governor Eyre's influence the legislature passed an act abolishing the constitution and virtually tendering the government to the crown. The act empowered her Majesty, the Queen, "to create and constitute a government for this island in such form and with such powers as to her Majesty may seem best fitting." This act, endorsed by the crown, was the final surrender of those liberties for which Jamaicans of other days had hotly contended : a representative government which had, in a history of two hundred and two years been almost Republican in its powers and pretentions.

Upon the report of Sir Henry Knight Stocks and his commission, upon the conduct of Governor Eyre, the crown refused to replace him at the head of the Jamaican government and he left Jamaica.

In 1865, the year of the insurrection, financial affairs were at their lowest ebb. In September, less than a month before the outbreak, the colonial treasurer showed a deficit of about £80,000, and this was followed by unusual expenses due to that affair. To cover these, a rum duty, house tax, and various tariff burdens were imposed. Trade licenses were required to be purchased by those engaged in certain branches of business. The result of these necessary enactments was a temporary revival of the treasury. Three years after Governor Eyre's departure there was a surplus of £5,599.

The year 1868 should be a red letter one in Jamaican annals. It was the turn of the tide, the dawning that came after the darkest night: the year of the first surplus ; the year of the first fruit shipment from Port Antonio ; of the revival of coolie immigration ; of the first cinchona planting on the Blue Mountain.

Sir Peter Grant was then governor. Throughout the whole of his administration of government there was an annual surplus in the treasury. Reporting on the financial situation in 1871–72 he says: " The continuing surplus accrues from no increase of taxation, and is in the face of a large expenditure on public works of utility and importance, of a largely increasing expenditure on such departments as those of education and medicine, and of some increase of expenditure in those administrative and revenue departments which necessarily require development as the population and wealth of the colony become developed." About the time that the report just quoted from was written, the import duty levied in the early part of 1868 was removed, and certain tonnage dues and taxes on live stock taken away.

1871 saw the disestablishment of the Church of England, the repeal of the law granting power to the governor to proclaim martial law in times of insurrection, and the taking of the census. The population was then estimated at 506,154. The seat of government was at this time removed from Spanish Town to Kingston, a move which was decidedly against the experience of nations, and could hardly be defended on the plea of convenience. Not only were the commodious buildings and government property abandoned and allowed to go to decay, but the defence of a retired position, the advantage of comparative isolation from the centre of business activity, and the value of historic association were alike given up for a position of small advantage to the routine of public work, whatever benefit it may be to the merchant or professional man.

Sir J. P. Grant had an opportunity to test the value of an island statute, relating to the confiscation of munitions of war landed in Jamaica.

The *La Have*, cleared for Kingston and loaded with arms, was captured by a Spanish man-of-war and brought to Jamaica, where the cargo was duly seized. The owners brought suit for £33,000 against the governor, who found himself so hard pushed that he was fain to compromise for £7,920, giving his note therefor. The colonial council redeemed the note and the Imperial Government finally refunded the money.

Sir William Gray superseded Sir J. P. Grant in 1874, and ruled till 1877. Though these years were disastrous in some respects, being marked by drought, floods, destruction of roads, and the small-pox, besides a financial crisis in which several prominent houses went under, yet there was also the establishment of the Kingston street cars and the completion of the Rio Cobre irrigation canal, a work of which it would be difficult to overestimate the value.

Lieutenant-Governor Rushworth succeeded to Sir William Gray in the management of the government, in 1877. Kingston was lighted with gas that year, Jamaica was admitted to the postal union, and the commission to inquire into the condition of the juvenile population appointed, with results decidedly beneficial, as it led to the establishment of the system of education now operative, besides calling attention to certain other necessary reforms. Before the end of the year the lieutenant-governor died, and Sir Anthony Musgrave succeeded him. At the beginning of his administration financial affairs were not in good shape. The transfer of a large immigration debt, together with hospital and other expenses, added to a deficit in the general account of £4,063 ; and an anticipated deficit for 1878 of £2,683 burdened the treasury. To meet the exigency the governor recommended that the poll tax on cattle, removed seven years before, should be reimposed and a loan raised. This was enacted, and thus began an administration which, while not always brilliantly successful financially, was still marked not only by the adoption of some neces-

sary expedients in raising the revenue, but by a generally wise and enlightened policy, and the institution of
a number of public works and reforms, by which the
island is still benefitted.

The expenditure on public works during the first
year enabled the treasurer to report a surplus, the new
loan being added to the public debt. But afterwards
the measures already alluded to were carried through
with judgment and vigor.

The railway and telegraph facilities now enjoyed by
the island are due to Governor Musgrave, as is also its
cable communication with the rest of the world. The
judicial system was improved and the consolidation of
the superior courts accomplished. The Victoria Institute, for the promotion of literature, science and art,
was established. The cinchona plantations were made
in St. Andrews, and an annual scholarship founded in
Kingston, which made possible to the holder admission to either of the English universities. Besides
these things a change in administration of the high
school and in the efficiency of the teachers' training
schools were inaugurated. In 1879 a new marriage law
was passed, making civil marriages legal.

Nature, during this administration, did not act as
the supporter of the governor and his council in their
efforts for the advancement of Jamaican interests.
Floods, a drought, a cyclone, earthquakes and other
calamities caused considerable distress, some loss of
life and injury to commerce. By wise management
much of the ill effect of these things was averted however.

The delay of the schooner *Florence*, laden with arms and bound for Venezuela, but driven into Kingston by stress of weather, caused complications which led to some debate and the recognition of all the non-official members of the council. The four parties to the disagreement were Sir Anthony Musgrave, and the Collector of Customs, who made the seizure; the Venezuelan owners of the cargo, the Secretary of State, who instructed the governor to request the legislative council to pay damages and costs after the suit, and the council, who claimed that the acts of the governor were entirely on the ground of international and imperial duty. This controversy led to the appointment of commissioners to inquire into the financial questions involved and to report upon them, who arrived just after the disastrous Kingston fire, which occurred in 1881, on the 11th of December. The commissioners adopted one important suggestion made by Sir Anthony Musgrave, that is, the creation of the office of Collector General, the incumbent of which should preside over the joint departments of the revenue and treasury  This was approved by the Secretary of State.

The Kingston fire, just referred to, swept over the town, damaging property to the value of £150,000. Great distress was occasioned, but without question the temporary loss and hurt were more than balanced by subsequent improvements in building and business.

The retirement of Governor Musgrave was the cause of sincere regret on the part of the people of

Jamaica, who recognized his wisdom and appreciated the earnestness of his efforts for their advantage.

During Governor Gamble's rule in 1883, promises were made by the crown that a constitutional change should be made for Jamaica, and the elective element, enjoyed for so many years prior to the disastrous events of Governor Eyre's administration, should be restored.

These promises were carried into effect in 1884, while Sir Henry Wylie Norman was at the head of affairs. A royal commission to report on the franchise consisted of the following named gentlemen: Hon. William Harriott Coke, Hon. William Vickers, Hon. Michael Solomon, Hon. Arthur Watson-Taylor, Hon. Thomas Lloyd Harvey, Mr. Samuel Constantine Burke, Dr. James Cecil Phillipo, and Mr. George Stiebel. Upon the reception of a despatch from Lord Derby, which did not promise to the people under the new constitution all the liberty of government which they sought, especially in questions of finance, there were numerous private and public meetings held in several parts of the island, and protests made against accepting anything less than a "definite, substantial and effective control over the financial affairs of the country." At a public meeting in Kingston a resolution was passed, protesting among other matters that "the governor's presence and power in the council has been in the past and will be in the future, unduly restrictive of the freedom of debate." To this the governor replied; the commissioners of finance commented upon it at some length and considerable warmth was shown in the controversy, but no conclusion was reached for a time. On June 20th of that year an order in council, by her Majesty, was issued, reconstituting the legislative council of Jamaica.

The elections were carried through with an absence of anything like excitement or disturbance; a quietude more surprising, when we consider that such important changes in the governmental affairs of other countries have frequently been accompanied by more or less disturbance and have led to unpleasant results.

Events more recent than those which we have so briefly noticed are not yet history, and wherever they come within the scope of this work must be treated in other chapters.

# CLIMATE.

THE varied surface of Jamaica, with altitudes ranging from the levels along the sea, up through the plateaus of the western end of the island to the 7,360 feet
of the Blue Mountain Peaks, affords a range of climate
which leaves little to be desired by either the seeker
after health or the permanent resident, provided the
ability to move from one elevation to the other is taken
for granted. It cannot be denied that during certain
seasons, principally from June 15 to September 15, life
in the larger towns is far from enjoyable on account of
the heat. Yet it is quite safe to say that during that
same period the resident of Kingston can maintain the
mental equipoise due to a freedom from overheating
with greater ease than can the sojourner in New York
City. This is due to several causes, principal among
which may be mentioned the following: the thermometer does not reach the "upper nineties;" there
is nearly always present during the midday a strong
sea breeze—commonly called "the Doctor;" life
goes on at a slower and more comfortable pace, the
houses, needing no provisions against the colds of fall
and winter, are constructed only with a view to keeping out the rain and the heat, and are, where well situated, perfect abodes of sombre and enchanting cool-

ness; and lastly, though principally, *the nights arc never unbearable.* On the outskirts of Kingston the thermometer, carefully watched during the hot month of August, 1890, by the Observer of the United States Signal Service, stationed there to warn his Government of approaching hurricanes, never registered above 88°, though the New York papers were at that time laden with complaints and accounts of the excessive heat throughout the Northern States, where even death was the result of exertion in a temperature ranging near the 100° mark. Just prior to this period one of the authors of this book was called away from Jamaica to New York. His letters to his more fortunate confrère, who remained behind in the upper Liguanea plain, St. Andrews, at an elevation of less than 500 feet, at the Constant Spring Hotel, were remarkable as coming from a denizen of the " bleak north-land " to a resident of the "sweltering tropics." While he was toiling over baked pavement with 98° in the shade as no unusual condition, and the humidity of the atmosphere a prominent characteristic, his collaberateur was not called on to withstand the effects of a daily average of over 80° and a maximum of 87°, in a remarkably dry atmosphere, and with nights when the mercury frequently went down to 63° and seldom remained over 70°. In this last particular the Jamaican climate is notably strong. Persons resident in the island for many years, have never experienced a time when during a whole night through, sleep was uncomfortable by reason of the heat. Rather is it likely to be disturbed by the necessity of procuring an extra blanket

or counterpane between the hours of two and five
A. M.

The temperature varies with the altitude: though
a rare occurrence, frost is to be found on these higher
peaks, and once in several years ice is said to form on
and near the Blue Mountain Peaks. The histories
speak of occasional visitations of cold winds through
these mountain districts, the climax seeming to have
been reached when, in 1671, was experienced "the
blast"—a severe and very cold wind which destroyed
many crops, not only in the hills but in the low
lands as well, where especially cacao and indigo
suffered.

While the general average of temperature is remarkably uniform throughout the island, the average rainfall presents phenomena which seem to be quite
beyond the present understanding of the student of
meteorology. The line of demarkation between two
adjoining districts is even more marked when considered from the point of rainfall, than is its natural line
perhaps formed by a precipitous mountain range. A
visitor to the Dry Harbor Mountains of St. Ann's may
find the inhabitants of Brownstown and vicinity actually suffering for water, and to a great extent depen-
dent upon the enterprise and liberal-mindedness of a
prominent doctor missionary who is the happy pos-
sessor of some famous tanks of rainwater, saved against
the day of need. Yet after a brief journey into the
adjoining parishes, both to southward and westward, he
will come to regions where the red clay and contrasting
deep greens will tell him of the almost daily heavy

showers, which render 'these plateaus at times' rather too damp for comfort.

Though along the highest ranges rains are of almost daily occurrence for most of the year, still for the island in general it is a usually safe rule to say, that the months of May and October witness the heights of the rainy seasons, and that from July to the beginning of the fall rainy season, usually late in September, heavy showers are very frequent. For two centuries this general rule has been observed to hold true. Sir Hans Sloane, the noted naturalist, writing of his experiences in Jamaica two hundred years ago, thus speaks in the preface to his " Natural History of Jamaica ":

" According to the different positions of the places, so the rains are more or less violent, and come at different times ; but, generally speaking, the two great rainy seasons are in May and October, in which months, at new or full moon, they begin, and continue day and night for a whole fortnight with great violence so that the earth in all level places is laid under water for some inches.     *     *     *     *     *

" In the month of January is likewise expected a season of rain, but this is not so constant nor violent as are the other two, and probably may come from the violent *norths* coming over the mountains with part of their rains with them ; for in the north side of the island rains in that month are generally very frequent and violent.     *     *     *     *

" For all the summer months, or when the sun is near or over their heads, or through almost the whole year, towards noon it rains on some part of the ridge of mountains, running through the island, with thunder and lightning. These rains seldom reach two or three miles into the plains ; wherefore, on account of these rains, any valleys lying very near or amongst the mountains have more seasons and are more fertile than the plains farther off, which, if they have any rain, is but the outskirts of that in the mountains and therefore inconsiderable."

Commenting on this very accurate account of the rainfall, Mr. Maxwell Hall, F. R. A. S., says :

" It thus appears that Sloane has alluded to the May and October rains, to the winter rains on the *north side*, to the summer rains on the central hills, and to the small rainfall on the southern plains. Consequently the characteristics of the rainfall have not altered for at least two hundred years."

It is to Mr. Hall that an intimate knowledge of the meteorological conditions holding throughout Jamaica is due. From Kempshot Observatory, about five miles northeast of Montego Bay and in latitude 18° 24′ 50″ 8 N., and W. longitude, 78° 52′ 22″ 8, at an altitude of 1,773 feet, Mr. Hall has for some years conducted and directed labors which have greatly added to the rather meagre tables which were in existence before he entered the field.

To compare the temperature of Jamaica with the northern States of the United States or with England, one should observe that the mean for the summer months reaches only about 81°, and that for the winter months it does not descend to 75°. With a maximum under 90° and a minimum over 70°, with a mean difference during the day and evening of 15.4°, life can never be an absolute burden, even in Kingston, where the severest conditions are to be found. When the fact that a fourteen mile ride will take the invalid above the fever line to an altitude of 4,000 feet, where the mean is fully 10° below the 78.6° of Kingston, is recalled, it will at once be seen that, to quote from the Hon. Geo. E. Hoskinson, late Consul for the United States at Kingston : " For people *of temperate habits,* Jamaica is as healthy a place for residence as any in the United States and in this I think the records of the Medical Bureau will bear me out."

The italics in the last sentence are ours. Nothing is more impressive to the visitor than the foolhardiness shown in this respect by many, especially young men, who, coming out from the colder north, usually during the winter months, when the contrast in temperature is very great, find here a social condition, among the better classes of men, which is famous for its hospitality and good-fellowship. The rum of the island, a most seductive beverage, lends itself readily to many decoctions which to the uninitiated are as injurious as they are novel.

While the native gentleman, though sometimes the happier, is very seldom indeed the worse for his glass, the stranger, on the other hand, though the last man to indulge in such freedom, too often attempts to vie with and outdo his hospitable acquaintance. The result is frequently an attack of " pernicious " fever, so called here, a form of fever, which though not by any means the dreaded "yellow jack," has no doubt often

been called upon to bear that fatal malady's burdens.
A concensus of opinion taken from numbers of the
medical men throughout the island, bears us out in
the statement that fully one-half the deaths of visitors
or temporary residents from febrile causes can readily
be traced to excesses in liquor or those exposures
which intoxication so generally leads to.

From the foregoing, however, it should not be
understood that deaths from these causes are common.
It is only the case that a death rate, already low, is
somewhat increased by these breaches of the laws of
hygiene.

In this connection it will interest the reader to know
that leading Life Insurance Companies in the United
States now recognize the fact that the same care taken
against overheating in Jamaica that it is presumed
their patrons take against the cold of the northern
winters will make it safe for them to allow their poli-
cies to cover residence in the island without restric-
tions or the increase of rates.

After what has been said of the varying altitudes
and the considerable range of temperature and humidity, it will at once be understood that Jamaica affords
rare opportunities for the health-seeker to obtain just
such conditions as are best suited to his case.  For a
full discussion of this subject the reader is referred to
a work entitled " The Climate of Jamaica," by James
Cecil Phillippo, M. D., L. R. C. S. (Edin.) etc., London,
T. & A. Churchill.  Dr. Phillippo has had many years'
experience as one of the foremost physicians in the
island.

In the "Handbook" for 1882, the Rev. Alexander Robb, D. D., says: "There are few of us but can tell of those we have known; seriously threatened and in danger, taking refuge here, with speedy and with much advantage. Persons who could scarcely have hoped to live through a northern winter, by coming hither have been so far restored as to live for years, and accomplished much important work.* * * In fact the experience is so extensive and its testimony is so assuring, that we may hold it settled, that persons of the class mentioned who can come to Jamaica may do so with the best of hopes."

Before closing this hasty consideration of Jamaica as a sanitarium, it is well to lay special stress on the fact that here life can be to a great extent out of doors, even to the advanced invalid. "Perpetual June," exactly describes the weather conditions to be found here during every one of the twelve months of the year. The Jamaica house is more like a series of closed verandas bunched together than any other form of domicile known to the northern builder. Here in these veranda-like rooms with the jalousie blinds closed in case of wind or rain, or with them thrown open to the free entrance of the balmy breezes most of the time, the invalid cannot fail to recover, unless perchance the exodus to Jamaica has been left until it is too late. Here the breeze is ozone laden to the fall; nature is ever in her brightest garb; and the cool nights insure the presence of the sweet restorer—sleep.

# COMMERCIAL LIFE AND RELATIONS.

"THE tone of thought in commercial circles in Kingston is now much more American than English; and reference is much more frequently made to the opinion of the States and New York than to that of England and London."

So writes an English author who has carefully considered his subject.

To begin: it must be remembered that Jamaica has practically no manufactories except those of sugar and rum, and a few products such as the bamboo, which an American at Black River is extracting the fibre from. The great bulk of what she ships must be either manufactured goods in transit or the natural products of her fields and groves. On the other hand it is apparent that most of the articles of personal or household use or adornment which the dweller in Jamaica uses, he must import from other countries.

Who are the parties most interested in this question of supply and demand?

Of the bananas shipped from Jamaica, out of 2,881,-313 bunches, 130 bunches went, in 1889, to the United Kingdom, and 2,879,560 to the United States. During the same time the United States took sugar to the value of £176,353, while England took £32,792 worth, and canada £27,632. Of rum (let our temperance readers note this) England received 1,216,012 gallons, while the United States contented themselves with 37,442 gallons and Canada got along with 5,823.

The total exports from the island during 1888–89 (which we quote because the official figures have not yet been published for '90 at the time this book goes to press) amounted in value to £1,614,824, of which nearly all is reckoned on island produce. Of this amount England paid £525,118, and the United States £793,310, after which came France, Germany, Canada in the order in which they are named.

There arrived from Great Britain during that last year 58 steamers and 17 sailing vessels with an aggregate tonnage (vessels in ballast only excluded) of 97,239; from the United States of America 276 steamers and 57 sail, the tonnage being 167,102. The vessels that cleared during this period registered as follows: To Great Britain 118,032 tons: to the United States 207,169 tons. For the past 10 years,

| IMPORTS. | EXPORTS. |
|---|---|
| 1879–80...............£1,475,197 | 1879–80............. £1,512,978 |
| 1880–81. ..............1,342,699 | 1880–81................1,178,594 |
| 1881–82. ... ......... .1,321,962 | 1881–82............. .1,549,058 |
| 1882–83..... ...........1,625,411 | 1882–83..... ..........1,469,446 |
| 1883–84...............1,568,639 | 1883–84..... ..........1,483,989 |
| 1884–85..... ...........1,487,833 | 1884–85.................1,408,848 |
| 1885–86...............1,325,603 | 1885–86............. .1,280,118 |
| 1886–87...............1,351,394 | 1886–87............ . .1,509,010 |
| 1887–88...............1,695,605 | 1887–88.................1,828,590 |
| 1888–89...............1,597,600 | 1888–89.................1,614,823 |

The imports were drawn as follows during the last two years:

|  | 1887–88. | 1888–89. |
|---|---|---|
| United Kingdom | 62.8 | 55. |
| United States | 27.7 | 33.9 |
| Canada | 6.6 | 9.8 |
| Other Countries | 2.9 | 1.3 |

This shows a marked increase made by the United States of America and Canada during one year. There was a decrease in most of the staples of exports over the maximum figures of the year before, but an increase over previous years. The trade with the United States is unquestionably growing.

It may surprise the Northerner—British or American—to learn that one of the largest trade correspondents, next to those we have been figuring upon, is the United States of Columbia.

There is a moral to all this. Whether Canada has understood it better than other countries and shown her appreciation of the situation by applying for more space at the coming exhibition, is a matter for consideration. Certainly the trade with an island containing 600,000 souls (and bodies as well) is an item well worth considering by any nation.

The recent publication of figures showing the increase of certain exports in July, 1890, over the corresponding month of the previous year, encourages us to believe that the advance is a steady and healthy one ; although in a few cases there has been a slight falling off in quantity, as is the case with cocoanuts. There has been an increase in the coffee export of nearly one half, and hides show about the same proportional advance. Sugar, fruit and rum have also increased. Bananas especially show a total of 445,512 bunches shipped to the United States and Canada as against 301,991 bunches in July, 1889.

Very lately a gentleman who has interested himself in oyster culture has been endeavoring to influence capital in that direction. He has taken the native Jamaican oyster north for examination and trial, and in return imports a number of the bivalves for transplanting in Caribbean water.

Manufacturers of Jamaica do not yet afford any appreciable part of her exports. The bamboo fibre works at Black River have made a beginning, which is mainly significant of what may be done with raw products in the island. In Manchester a saw mill company has been recently established, capable of turning out 5,000 feet of boards per diem. A very small amount of shipbuilding is carried on, principally at St. Ann's Bay.

The staple manufactures are sugar and rum. The crushing of the cane and extraction of sap from which the sugar is made, and the refuse of which furnishes the material for rum distillation, are carried on everywhere throughout the island. From the primitive bamboo frame, mill and mule walk beside some mountain hut, to the "Wetzel," "Aspinwall" or "vacuum" apparatus and centrifugal drying process in use on the great estates, the joint production of sweet and strong goes on. Jamaica's export of these two products aggregates between £300,000 and £400,000 a year in value.

The working day for outdoor laborers is supposed to be ten hours, in the neighborhood of Kingston, and eight hours in the country. Mechanics get from 2*s*. 3*d*. to 5*s*.6*d*. a day ; male laborers 1*s*. 6*d*. to 2*s*. and women 9*d*. to 1*s*.  A team of two mules, with driver, costs 7*s*. per diem.

But much of the work, especially in the country, is done by what is known as " task work," evidently a survival of the slave days ; the prices for which are generally low, except for certain branches of building and masonry work.  There is a scarcity of skilled labor.  The sugar estates find common laborers scarce and the government works are over supplied.

Jamaica consumes a great deal of material produced and manufactured in other countries.  As already shown she receives most from Great Britain and sends most to the United States.  Nevertheless it is easy to find all of the necessaries and most of the luxuries of modern life, American as well as English, at moderate prices in most of the principal towns.  The food supply is ample and cheap, fruit being especially so.  In general, prices compare favorably with those of northern countries, even for imported goods.

# THE INDUSTRIAL EXHIBITION.

In September, 1889, Mr. William Fawcett brought before the board of directors of the Jamaica Institute a proposition "for holding an exhibition in Jamaica, illustrative of its natural products and their manufacture, combined with a loan art exhibition." A plan, elaborated from Mr. Fawcett's report, was prepared by the board and laid before the governor. His Excellency entered heartily into the plan and has given it his unstinted support from that hour to this, leaving no stone unturned to insure its entire success.

A meeting of Kingston gentlemen with the governor, in the library, was held on the 19th of September; the subject was thoroughly discussed at that time, and the pledges of support made by those present gave evidence of the faith which representative Jamaicans have in their island's capabilities and progress.

The following resolutions were agreed to at that meeting: "1st. That the other West Indian colonies be invited to send such exhibits as will clearly indicate the great resources of these colonies. 2d. That special exhibits be invited from England and other countries with which we trade. 3d. That, in order to provide the necessary funds for carrying out the project in a thoroughly effectual manner, gentlemen of the island be asked to become guarantors to the extent of £10 each and upwards."

Working committees were soon appointed by the executive committee, an act of incorporation was passed, and commissioners, appointed by the governor, began to make preparations for the event.

At the commissioners' first meeting Quebec Lodge was recommended as a site by the building committee. A set of plans and estimates was also submitted, the whole to cost above £14,000.

The site of the building (Quebec Lodge) is to the north of the race-course, a mile and a half from Kingston harbor and two hundred feet above the sea.

The building is of Moorish architecture, the central nave 511 × 40 feet, and side aisles making the width 81 feet. The transepts are 174 feet long and the same breadth as nave and aisles. The height of the dome is 114 and the minarets 74 feet. It is a wooden building, well protected, and provided with escapes for fire, etc. The long balconies will be used as promenades. The main entrance is at the east end, and a carriage way at the south. The grounds are handsomely laid out with ornamental walks, gardens and buildings. A band stand, concert hall, Jamaica village, and separate exhibits will add to the attractiveness of the exhibition.

For months the Half-way Tree road has been the scene of the transportation of giant palm trees and other plants. which, *en route* to the exhibition grounds, are being drawn by patient ox-teams. The display of

tropical trees and plants will be one of the great features of the exhibition for the visitor from the North.

The guarantee fund at the end of June, was officially
reported to be £27,079 10*s.* o*d.* Of this the largest
amount came from Kingston and the next from St.
Catherine, the smallest being from Hanover and
Trelawney.

Conversing with the guarantors, there seems to be a
general impression among them that they will lose the
money subscribed, yet we have not found one who re-
grets lending a hand, or who doubts the ultimate
benefit that Jamaica will receive from the fair.

The advance that the banks refused to make in
December of last year (1889) private parties supplied,
with that generosity which has been characteristic of
those interested in this enterprise. Mr. Stiebel loaned
£5000, Colonel Ward followed suit with a similar sum,
and Mr. Verley completed the £15,000.

The government sanctioned an advance of another
£15,000 from the public treasury. This £30,000, the
estimated cost, was secured, £25,000 at 3%, and the
remainder at 6%.

Committees in various countries have been charged
with the work of forwarding the interests of the
exhibition. In Canada the committee has done very
good and efficient work, if we may judge by the results. The people of the United States have never
been averse to entering largely into such an enterprise
as this when it is properly represented. Yet the
Jamaica exhibition will be almost entirely without a
showing from the United States.

All exhibits will be arranged under the following groups.

Group 1. Raw Material.

Group 2. Implements for obtaining Raw Material.

Group 3. Machines and processes used in preparing and making up the Raw Material into finished products.

Group 4. Manufactured goods.

Group 5. Education.

Group 6. Fine arts, literature, science.

Motive power to the extent of 100 horse-power will be provided by the commissioners, free of charge : but all counter shafting, pulleys, and connection with the main steam pipe must by provided by exhibitors. The exhibition will be open to the public on the 27th of January, 1891.

There was at first a great deal of misunderstanding among the peasantry as to the object of the exhibition. A common belief in some quarters was that the government wants to find out what they possess in order to tax them more heavily. Poor souls ! many of them have not yet learned what the government is trying nobly to teach them, that power and benificence may go hand in hand. Owing to the strenuous efforts being made by his Excellency and his supporters, the lesson is, we trust, gradually being learned.

The application of Austria for four thousand square feet of space for exhibits, led to the serious consideration of the question of new buildings. The result of this has been the erection of additional annexes. There are now several structures besides the main one already described. A pavilion for the special exhibits under the direction of one large firm is noticeable for its beauty. An amusement hall, erected for theatrical performances to be furnished by an English company, and capable of seating a large audience is also a recent feature.

A large annex for general exhibits was decided upon in September, its dimensions to be 400 by 40 feet. The machinery hall measures 100 by 50 feet, and its cost is £500. All the annexes as well as the main buildings, have been provided with electric lighting apparatus.

The available space in the main building is 40,000 square feet. As England required at least 5000 feet of this; Canada 8000; Austria 4000 and other countries 15,000, and Jamaica runs the amount of space required up to 50,000 square feet, exclusive of 20,000 feet of walking space, the necessity for the annexes will be apparent.

In the extensive grounds the visitor will find an electric railway, a toboggan slide, merry go round, etc. An artificial lake with "leviathan" for passengers has been furnished by Col. Forster of New York. The extension of the grounds, which makes a further encroachment upon the forty-five acres of the race grounds, renders these additional attractions and displays possible.

To the delight of many who thought the outdoor space at first allotted inadequate, the acquisition of additional ground was also made early in the autumn. This of course gave opportunity for an increased floral and arboreal display. Although not completed when this work goes to press, enough has been done to convince those who have had an opportunity to note the progress of this garden that it will equal in tropical beauty and luxuriance, anything that the world has produced in modern times.

The Austrian exhibit, we are informed, will be arranged by artists from Europe, sent especially to attend to this work.

England's grant of £10,000 has been a great benefit to the exhibition. The refusal of the United States to show any interest in the matter is one of those evidences of want of foresight that is to be deplored. Whatever has been done by American exhibitors has been without the encouragement of their government.

Among the curious exhibits recently noted has been one by the military department and another by the English Post Office Department. The West India Islands have been well represented and in fact there has been no lack of effort to push this great undertaking to a successful conclusion.

The local shows, which in Clarendon, Portland, Hanover and elsewhere have aided greatly in awakening interest and enthusiasm in the people of the island, have proven by the result the wisdom of those who urged them on. The Clarendon Exhibition or "show" held at May Pen, was quite successful. The river, which was dry when we last saw it, had considerable water flowing in the middle of September and this added to the attractiveness of the place for visitors.

After considerable debate about the Portland fair, at Port Antonio, it was decided to hold it in the fort.

What the outcome of Jamaica's great industrial exhibition will be remains to be shown. We have little doubt of its success. It has already awakened some who were in a state of lethargy, and has even now begun to rouse a spirit of emulation which cannot fail to be beneficial to the people at large.

# A CITY UNDER THE SUN.

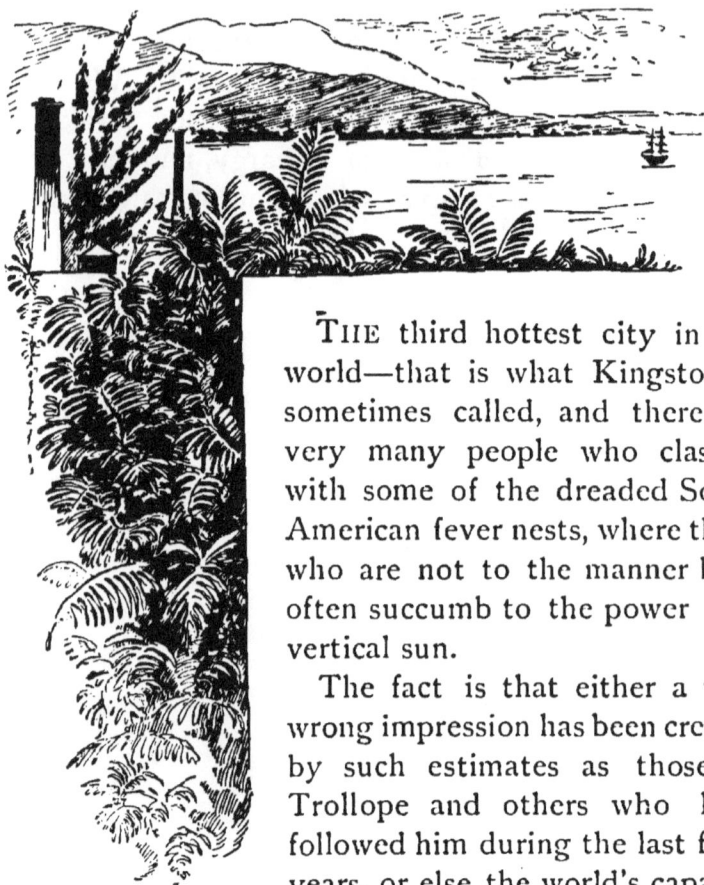

THE third hottest city in the world—that is what Kingston is sometimes called, and there are very many people who class it with some of the dreaded South American fever nests, where those who are not to the manner born often succumb to the power of a vertical sun.

The fact is that either a very wrong impression has been created by such estimates as those of Trollope and others who have followed him during the last forty years, or else the world's capacity to provide hot places has been hugely overestimated, for certainly a large and contented foreign population make Kingston their temporary habitat, and, with a little attention to the commonest hygienic laws, are not only able to exist, but to be comfortable.

Another false impression given is that the streets are all bad; and a third, that the rainy season is apt to overtake the unsuspecting Northern foreigner with diluvian intensity, and make the Jamaican metropolis a repetition of Venice.

The fact is that Kingston may have sixty showers in the course of a twelvemonth, a large percentage of the annual rainfall having been known to occur in a very few hours. While the showers last, owing to the backing of high hills which the city has, the flood descends as soon as the rains fall, and the immediate and pleasant result is that the drains and gutters are thus flushed, to the preservation of health and the conservation of comfort.

Indeed, a wise and careful Providence has sent three agents to purify this old city and make it comfortably habitable; these are the rains, the Doctor, and John Crow. By the Doctor the Jamaican means the breeze that pays a morning and an evening visit, laden with comfort and life. John Crow is the first being to attract the attention of the tourist, even before he falls into the clutches of those fiends, the cabmen, at the wharf. John has a leisurely, familiar way with him. If life is worth living, it is at least not worth hurrying for, in his estimation. Black of coat, ragged of wing, red as to head, and heavy in flight, besides being misnamed "*vulture*," poor John Crow is pro-

tected by law as a useful scavenger. Sometimes
people call him the street-cleaner. He is a member of
the buzzard connection, but with a lurid head and ruff
of feathers encircling his bare neck that suggest the
vulture. No one has taken the census of the John
Crow tribe, but it is safe to assume that their number
reaches far into the thousands. It is almost impossible to look upward without seeing several of them at
once ; they come down within a few feet of one, if
there is any inducement offered, and roost in the trees
about a house at night, or do their dreaming upon
your fence or ridge-pole.

At first the Northerner receives a multitude of im-
pressions that clamor to be assorted and classified.
He is amused, interested, perhaps disgusted, with the
extortionate cabman who sees in every stranger a possible victim. But if the new-comer falls as we did, into
such hands as those of Old Joe, his wounded feelings
are speedily soothed and his mind set at rest. There
are servants and servants in Kingston as elsewhere.
Among servants, Old Joe stands pre-eminent. His
kind, wrinkled face is wreathed in a perpetual smile.
His aged hands are clasped and unclasped as he bows
slightly over them while he talks. There is comfort in
his attention at the table, and consolation to the home-
sick, all-alone traveller in the way he enters your room
after the boots in the morning, and " hopes Maastah is
well this maanin'." We have seldom heard the broad
"a" of the local dialect sound so musically in any
one's mouth.

Do you imagine Joe is a myth? For a truth we

THE CITY OF KINGSTON.

wish all myths of the better kind were as real and true.
No one has engendered more of a comforting faith in
humanity for many and many a day than this kind old
ignorant brown man. Did he not see that A was
lonely and try to amuse him ; that B was curious and
grateful for enlightenment on many points, and that
poor P. (having imprudently challenged the midday sun
in heavy clothing and black hat) had caught a fever and
needed nursing? Day after day, when the patient
in his weakness and loneliness was calling for the ab-
sent wife, the faithful old servitor found time from his
multiform duties to wait on and watch him. Night
after night did he not sleep within call on the floor of
the piazza at the threshold of the sick-room?

When you come to Jamaica you must find Joe.

There is constant variety and surprise in the negro
character. There was John Williams, for instance,
whose cheerful salutation, " I come, sah ! " was as en-
livening as his smile. But, in the language of one of
his companions, John Williams " done a crime, sah;"
and we saw him no more. His last " I gone " was
prophetic. We missed the cheerful, ingenious chatter
and barbarous dialect with which he regaled us as he
sat, or rather reclined, on the floor, with his bundle of
papers beside him.

Kingston in itself is a city of moderate attractions
to one who is accustomed to tropical ways and growths.
Of course the tourist who comes here first cannot fail
to be impressed by the vegetation, many fruit trees
showing their heads over the houses, and the palm
towering everywhere with its long, branch-like leaves

incessantly moving. He must be interested in buildings very different from anything he has seen before,
and manners and customs so strikingly novel. The
women working the streets with hoe or shovel; the
lazy negro messengers, bearing whatever burdens they
may have on their heads; the black policeman, the
pretty Creole girl, the wrinkled venders of fruits or
sweets who are lingering in the cool, distinct shadow
which some porch paints on the white, glaring street—
all appeal to us as types of an unfamiliar life.

Yonder comes a brown soldier with turban, tight
jacket, and Zouavē rigging aft. Near him is a coolie
woman, who is gorgeously apparelled, her small head
decorated with gaudy kerchief and ornaments of silver,
her lithe body wrapped in parti-colored garments, broad
bracelets of silver and anklets of the same upon her
bare arms and brown ankles.

Then one listens with interest to the various street
cries, each one ending with "gwine by;"which is doubtless a reminder that all things in this world are but tran-
sitory after all. A clergyman could preach quite a sermon on "gwine by." Or what a splendid chance for the
temperance lecturer to take his text from the call that
rouses him in the morning—"Wi' pi', chapai pi, whiskey
bot'l gwi' by." Could you guess that, when translated,
this means—"Wine pint, champagne pint, whiskey bot-
tle; going by," and that the enigma is uttered by a
woman whose business is the collecting of bottles?

Hoping to find something especially characteristic,
a small party visited the Roman Catholic church one
evening when the bishop was announced to preach.

The auditorium was well filled, for though Jamaica is not a Catholic country there are a number of that sect. The cathedral is rather a fine one, and the altar, rich with color and brilliant with the light of many candles, made a fine background for the acolytes in scarlet and white, or the priests decked in the emblems of clerical rank. After considerable preliminary exercises, which the uninitiated stranger could not be expected to understand, and an antiphony which was more attractive, presently there was a great swinging of censers, and his grace the bishop appeared, two pages holding the hem of his garment.

Soon followed the surprise of the evening: a plain, practical, common-sense talk, worded so that the humblest of his hearers could understand it, and full of a most tolerant spirit. In the course of it, while dwelling upon those virtues—especially charity—which the preacher believed to be pre-eminently characteristic of the Catholic Church, he distinctly stated that he did not know how far they were carried out in other churches, and said that the idea of being angry with another because of difference of belief was irrational and wicked.

In making a tour of the city's buildings and points of interest, one is attracted sooner or later to the museum and library, the latter containing at present about twelve thousand volumes, among which are a number of rare old books and pamphlets upon the history, geography, natural history, botany, etc., etc., of Jamaica. Jamaica-ana is not a pretty-looking word, but that is what we mean. The library is rich in that sort of thing, and the obliging custodians of these treasures are very ready to assist the delver after old records. And what a field is here presented! We go back to the days of Spanish rule, of piratic atrocity, of English occupancy, and of slave insurrection. Penn and Venables, Morgan, the greatest pirate that ever lived and the great earthquake that destroyed Port Royal in a moment, all seem to be brought to our very time.

Down in the museum is a gruesome relic that was dug up nearly a generation ago. It is a cage of strap iron, so constructed as to fit the human body, with bands around the neck, breast and loins, bars and stir-rups for the legs and feet (the last having sharp spikes to press into the soles of the occupant's feet), and a ring at the top of the structure to suspend it by. This awful instrument of a fiendish tyranny contained, when exhumed, the bones of a woman, who had without doubt thus satisfied the malice of an enemy, or, more likely still, the barbarous passion of a master to whom the life of a slave was a thing of less value than that of his dog.

Kingston was originally built in the form of a cross. King Street, running north and south, crossed Queen Street, which is laid east and west. At their intersection is the "parade ground," a pleasant little park with trees and fountain, tastefully arranged walks and flower gardens; a favorite resting-place for the people after the heat of the day is done. King and Queen Streets are each sixty-six feet wide. The town has grown since they were laid out, to the northward and eastward of the Plaza.

Kingston is not only the metropolis and capital of Jamaica, it is also the most important city in the British West Indies. Its population number individuals of nearly every race, English and Americans predominating among the whites. It was built, as before stated, immediately after the destruction of Port Royal. For years the city possessed a charter and a seal, but surrendered them when the ancient rights of the island were given up in Governor Eyre's time. It has been visited during its history by four great fires. The first, in 1780, caused a loss estimated at £30,000. The second, in 1843, swept the city from the east end of Harbour Street to the Catholic chapel at the end of Duke Street. The third fire occurred in 1862, and burned down stores, wharves and other property valued at £90,000. The fourth was in 1882, a disaster still fresh in the minds of residents, who lost their all there. Six thousand people were rendered homeless by this conflagration, and a large portion of the business part of the town burned over. A great many, even of those who suffered at the time, consider now that the fire was a blessing, since the rebuilt streets are finer and .more substantial than they were previously.

The city no longer draws its water supply from the precarious source of wells and cisterns, as formerly. About 1848 a private company brought water from the Hope River, and to this supply has since been added that from the Wag Water. The pressure is sufficient for all fire purposes and the system of filtration used results in a supply for drinking purposes that probably no tropical city in the world can excel, and few can equal.

In the year 1872, the seat of government was removed from Spanish Town to Kingston : a move the wisdom of which we venture to doubt, since the contingencies of war, riot, or conflagration would first menace the larger city, interfere with the public business and endanger the official records.

At the last census the population numbered between thirty-five and forty thousand people. Several building societies do a large business. Banks, life and fire assurance companies and discount associations flourish. Ice is manufactured and sold at the uniform price of seventy cents per hundred pounds, and electric lights are beginning to take the place of gas in the principal buildings and will soon supplant it in the streets. Street cars, drawn by mules, traverse the principal thoroughfares.

Among Kingston's buildings, the finest are the Victoria Market, at the foot of King Street ; the hospital, on North Street ; the Colonial Bank, on Duke Street ; the old parish church, on King Street, near the Parade, in which Admiral Benbow is buried, and where half the historical events of the last two centuries centre ; the colonial secretary's office, and that group of buildings between Church and East Streets, and the library and museum, to which reference has already been made elsewhere. The new building erected for the exhibition will be described more fully in a separate chapter. It is erected on the race course grounds, beyond the city limits.

The principal approaches to Kingston are by the Jamaica railway and the Spanish Town road on the west; the Halfway-Tree road on the north side, and the Windward road on the east. It fronts on a fine harbor. The lines of the Jamaica Tramway Company are laid from the foot of King Street to Constant Spring, by way of Orange Street, the Slipe Pen Road and Halfway-Tree; to the top of East Street; to Paradise Street on the Windward Road; to the General Penitentiary at Rae Town; to the Jamaica Railway, and to the May Pen Cemetery on the road to Spanish Town; and cars run on the lines at intervals of twenty minutes.

There is a fine statue to Sir Charles Metcalfe standing at the King Street entrance to the Parade. The Falmouth Court-house possesses a full length portrait of the same estimable governor. The only difficulty with these works of art is that they do not resemble each other. Other statues are those of the Hon. Edward Jordon, C. B., and Dr. Lewis Q. Bowerbank. In their anxiety to protect these works of art from the vandalism of advertisers, the authorities have caused notices to be placed on each one in which the words " Bill Stickers," appear in such large type that they are noticeable at quite a distance. Now it happens that in the city of Boston, some years ago there was a notorious character by the name of Bill Stickers. The story is told that not long ago, a northern visitor stopped in front of Sir Charles' effigy and exclaimed, " I knew that Bill Stickers was dead, but I never supposed they would erect a monument over him."

A word about Kingston shops, and that means Jamaican shops generally. A few have fixed prices ; the many have not yet arrived at that pitch of mercantile perfection, and the stranger should be wary about purchasing of a house he does not know, until he has learned enough of values to enable him to estimate for himself the worth of what he would buy. Clothing is generally much cheaper than in America. Books cost less, usually. Food supplies are cheaper, with the exception of meats and poultry, which are nearly the same. Fruits are especially low in price. Generally the cost of living is not so great as in the North in the most favorable season of the year. The rates for 'bus hire or car fare within city limits are 6*d.* One is besieged by beggars and small boys, who proffer all sorts of service for " quattie," or one-quarter of 6*d.* Labor wages are lower than those of the United States, and higher than in England.

Kingston has about a dozen lodging houses and as many taverns, where board and lodging may be had at rates varying from 35*s.* to 60*s.* per week, or from £1 to £2 2*s.* a day. The new hotel, erected in the site of the old Myrtle Bank, on the south side of Harbour Street ; and Park Lodge Hotel, in Burketown, on East Queen Street, are among the best of Kingston's hostelries. The first is American, and the second Creole in style.

There are places of worship for those of Roman Catholic, Presbyterian, Wesleyan, Baptist or Hebrew faith, besides the Church of England, which here, as in the mother country, assumes first rank.

# PORT ROYAL.

AROUND the wharves of Kingston are a number of small craft, sloop-rigged, with great length of boom and no great height of mast. They are not much to look at, but are none the less loudly lauded by the black men whose business it is to advertise them and solicit patronage. In spite of their dingy appearance these little vessels bear generally the names of royal personages, and we are vociferously informed that the *Queen* will take us safely, or we will find the *Princess* pleasant, or else be assured that the *Marquis* has first claim to our attention because of his bounding speed. We feel that the staid respectability of the *Queen* more nearly suits our requirements than the giddy allurements of her younger rivals, and are soon following a stalwart negro who is her proud " cap'n."

We know of no piece of water of equal size that presents as much food for meditation as this old harbor. Here lay the fleets of early Spanish explorers and navigators. Here were anchored the squadron of Penn and Venables, whose followers gave Jamaica to the English in Cromwell's day. It was here, too, that the most noted pirates and buccaneers the world has ever known, Morgan and Bartholomew, and others of that ilk, brought their booty after the conquest of Spanish galleons and of South American cities. Here were landed the spoils of Panama, the ransom of Maracaibo and Gibralta, or the gold and jewels and silks wrested from rich merchantmen bound to His-paniola. How often have the waters of Kingston harbor borne the sounds of unholy revelry or the hills around echoed the sterner notes of strife. It was beneath these waves that a squadron of England's war ships found a grave during one of those terrible and sudden hurricanes that devastate these latitudes. The fast growing coral covers them and the more pacific hulls of trading vessels that have from time to time met the same untoward fate.

But of all the grim stories that the night winds whis-per, the weirdest is that of the lost city that went down in the twinkling of an eye, with her young men and maidens, old men and children : with the wine of the feaster half drunk and the prayer of remorse half said ; with unfinished curse, uncompleted crime, arrested cruelty, in all its splendor and guilt.

So dreaming and moralizing and balancing in our
little boat, whose chief merit seemed to be that she for-
got to upset when her boom trailed in the water, we
and our freight of cameras and sketch books were
ferried over to the old metropolis of the West Indies.
There was little of it that we could visit, for before we
landed we had passed the walls and streets, the spires,
brothels and market places of the olden time. Many
fathoms beneath our boat's keel we knew they lay,
with sea-weed growing in the gardens and coral on
the walls, beneath whose fretted roof the shark lurked
in the halls where men had feasted, in the chambers
where maidens had slept and now sleep untroubled till
the last trump shall wake them.

After landing and threading our way through a succession of narrow streets, through groups of dirty
women with smiling faces and clothes of more than
Grecian simplicity, through companies of staring
children, of white turbaned, baggy-trousered colored
soldiers, or of predatory goats, we skirted the barracks
and the high walls which enclosed the parade ground
and entered the most interesting edifice we could find.
It was a church, old and inviting, but not nearly as
old as its reputation claimed for it; for we had been
told with all the prompt and cheerful inaccuracy
which is characteristic of Jamaica, that it had been
built before the earthquake.

Of course we know that the great Catholic church
of old Port Royal was somewhere down at the bottom
of the harbor, but it was possible that this smaller
house of worship might have been also one of the

ancient landmarks. It was a hope soon to be dis-
pelled, however. Upon entering, one of the first
things to meet the eye was a marble tablet on the wall
at the right of the door. This tablet was framed by a
pair of remarkable black dolphins done in some sort of
cement, and the old-fashioned letters upon it chron-
icled that the church was built under the direction
of two church wardens whose names were then
immortalized in the years 1725–26. That settled the
question of anti-earthquake construction, but it could
not spoil the intrinsic charm of the old-fashioned altar
or the delightful quaintness of the great mahogany
piece of furniture that, adorned with candles and sup-
plemented by a winding stairway, served as an organ
loft.

But perhaps the most fascinating and certainly the
saddest things in that old church, were the constantly
recurring mural tablets recording the virtues of the
many dead of the English communion, whose last
breaths had been drawn far away from home. Sometimes sacred to the memory of one, sometimes of an
entire crew; this one erected by the affection of
a sister, and that by the piety of a comrade; four-
fifths told of victims to that dreaded scourge, yellow
fever. Judging by these records it would seem that
in the old days Port Royal was a graveyard for the
British naval officers and seamen, a very plague spot
where the hardiest laid down his life.

Well, facts are about the last things that the world
learns, and a surface indication of something, especially
if that something is an unpleasant one, will generally

usurp the place of valuable evidence. The ancient metropolis was undoubtedly the port where many of Great Britain's sailors were permanently discharged— were mustered out of the service by that grim old officer, death. But it was also true that for years Great Britain had *no other marine hospital* in that part of the world than the one at Port Royal, so that officers and men from infected ports and vessels in South America, Central America and the Antilles, were all brought to Jamaica to die.

Vessels that had never visited Jamaica sent their crews thither by other vessels, and the result was an importation of disease that in most tropical countries would have proved far more disastrous to the country at large. Still, these tablets shock the visitor.

Though we know that things are changed now, and that the sanitary conditions of Port Royal are so greatly improved that there is hardly a possibility of a return to the old scourge, yet one cannot avoid a feeling of chill and fear almost, as he sees these dreadful reminders of the reign of the yellow death, three and four deep, covering almost the entire wall space of this house of worship.

There was a woman in the church cleaning it, and with her a little round-faced, big-eyed, bandy-legged girl of about two or three years, who looked at us solemnly, and then pattered away over the tiled floor to the protection of her mother.

When we left this place we went to the lookout, and getting from the negress in charge of the premises a statement that, " De hofficer in chaage ain' heah, saah, but I doan' t'ink he objec' to you goin' up," we ascended, taking care to leave the camera below, but smuggling a sketch book. The view from the top of that tower was well worth sketching. There were not only the military works of Port Royal, and the

more or less modern houses that we tried in vain to think might cover the six thousand population that is claimed for this ancient town, but beyond them we could also see the forts and harbor defences, the town of Kingston across the bay, and the velvet green mountains, with their violet shadows, always capped with clouds, beyond them. After a visit to the beach, and the exploration of one or two of the older corners of the town, we found that we had reached our time limit, and must find our way again to the ferry.

This time another member of the royal family took us in. Our Charon was quite a youth, and his face was so utterly expressionless that we doubted whether a cyclone or an earthquake could ever put life into it. But he managed to get us safely across, though the wind blew and the waters rose so that our vessel became a sort of bath-tub where we enjoyed (?) a perpetual *douche*, without the luxury of towels. Nor were we the only passengers. On the contrary, our boat carried a dozen voyagers of various shades of black and brown, most of whom showed unmistakable signs of fright before we arrived safely at the Kingston wharf.

The fact that two boats were overset in the blow that afternoon, and one poor fellow drowned, gave the last part of our visit to Jamaica's old metropolis something of the dignity of an adventure.

# ALONG THE RAILWAY.

WE are accustomed to railways that rush four-in-hand along northern river banks, or burrow shrieking into the mountain sides, or span cañons, or traverse the almost limitless plains where the sage and the chaparral flourish. We climb sierras that way now, cross valleys and watercourses, explore wildernesses, and there seems nothing strange or unnatural in seeing the brightest sky stained by the locomotive smoke rings, or hearing the deepest solitude disturbed by its strident voice.

But here is a railway that pursues the even tenor of its way between groves of mangoes, cactus hedges, logwood copses and banana walks; steals into vistas framed by great silk cotton trees and winds by the edge of streams over which the cocoanuts lean and beside which the pineapples grow. It mounts foot by foot to the higher island levels, around the heads of glens where the strange trees stand deep in ferns, and crowned with bright blossoms and gay moss streamers.

There is nothing in the world that indicates progress in industrial matters more than a successful railway. Every puff of an engine that is drawing its share of a steady output of produce, talks in a language that is intelligible to the dullest man. In the catechism of mechanics the question, "What is the chief end of a railway?" should be answered thus: "The chief end of a railway is to develop the country through which it passes; to make land valuable; increase commerce, and contrive, by such honest means, to afford a dividend."

At the Kingston station one is struck with the unique character of the place and the people; especially the latter. The cars drawn up to the platform, are built upon the same pattern as those of any English railway. They are divided into transverse apartments, which are entered from the side, having no connection with each other. The guards here are polite colored men, in a military-looking uniform. In place of the various phases of British or American life, we find a heterogeneous assortment of humanity with greater contrasts of color, character, creed and costume.

The Creole of position with his visiting cousin from Europe, or the American continent, takes his place in a first-class carriage. There is apt to be a pretty girl or two in his party, but they are for the most part, too demure to notice the stranger who is trying to use his eyes to the best advantage. There are colored people, black people, white people; there are faces that show Castilian origin, others of a Caledonian cast, many that are browned by more than exposure to a tropic sun.

Here is the bare-armed, braceletted, long-haired, coolie woman, with her babe partly wrapped in the gaudy shawl that is thrown half around the mother's head, half over her shoulder. Her wealth is apparent to all eyes, for she carries it where all may see, displayed upon wrists, forehead, breast, ankles; in fact anywhere that there is a chance to place a hoop or a bangle. Then there are others: soldiers, dressed in white and red uniforms; negro market women, bundle topped; newsboys and porters. These all speak a various language, in which they have a Babel proficiency. It pretends to be English; it sounds as though " Jabberwocky " had found a local habitation.

We do not believe there is a case on record where the stranger has been able at once to understand the English of the Jamaican of the lower class.

The train-shed into which the station building proper opens is about three hundred feet long, and wide enough to admit several trains abreast. Beyond this structure are the shops, engine houses, etc.

After passing the purlieus of the city, among the first scenes to attract attention are the extensive stock yards of Cumberland Pen, one of the large properties of the island, where great herds of horses and cattle

are bred and grazed. Cumberland Pen embraces a
good race-course, and its turf events are always looked
forward to with considerable interest, by both Creoles
and aliens.

The grazing pastures as seen from the railway,
present a clean, well-trimmed appearance. As high as
a cow can reach the leaves of the mangoes are close
cropped as though by a machine. One could sight
along the under sides of them as along the level of a
ceiling. Beneath, the grass is as smoothly trimmed.
Cows, and indeed all hoof-kind, are great conservers of
park and lawn. At another place, where the road
crosses some little stream or canal, a widening pool by
the wayside is literally full of horses, playing or standing shoulder deep, enjoying that advantage gained
over the flies. So we pass from point to point of
interest, here stopping at some way station and anon
skirting a sugar plantation beyond whose levels of
corn green cane the picturesque buildings, mills and
aqueducts cluster, and along whose borders the rows
of bamboos stand sentinel.

There is something which strikes one with surprise
in every phase of Jamaican industry, and this element
is not wanting here. As the water comes from the
aqueduct, it is distributed in channels or trenches, and
we have seen a number of women sitting by the side
of one of these little canals, throwing the water with
their hands upon the cane rows. At one point, a very
fine aqueduct, built of stone and supported on numberless arches, crosses the railway. It is picturesquely
draped with moss vines and ferns. The water it con-

ducts is drawn from the Rio Cobre, the beautiful river that waters the plain of St. Catherine.

Half way between Kingston and Spanish Town we get glimpses of the harbor head and the distant Healthshire Hills; or, as the old books call them, "Hellshire Hills."

One is continually struck with the indications of natural richness and fertility of the country, and the little advantage that is taken of these means of wealth by the inhabitants. It is probably true that the country is underpopulated. It is certainly a fact that every mile of it calls loudly for more intelligent methods of labor, and more earnest purpose in utilizing the natural resources. On the lower end of the line, the estates and fruit cultivations are made more valuable by the Rio Cobre canal, a public work accomplished several years ago, which has added thousands of pounds in value to the lands through which its irrigating stream passes. But beyond this there seems to be very little done to add to those natural gifts of soil and climate which we have more than once dwelt upon. From the very beginning to the conclusion of planting and gathering, the Jamaican cultivator rests in the knowledge that it would be a difficult matter to starve him out. One potent reason for this apparent apathy is a real want of capital. The man who clears his woodland with machete and fire, may be alive to the value of stump pulling machinery yet find it beyond his means. But many of the smaller cultivators are perfectly content to work according to old methods. Their oranges, they say, are the best in the world; why seek by grafting or budding, to improve them? Acres run to waste, and their owners wonder that foreigners doubt that they are making the most of their opportunities. Well, perhaps they are right. Why should they not rest, since nature is so willing to assume all the responsibility of crops and harvests?

Among the various products which the Jamaica Railway has made marketable one of the most valuable is logwood. At many of the stations we see great piles of the sticks, or of crooked roots, ready for shipment. Car after car passes us, loaded with this same wealth. The logwood grows wild, thousands of acres

of it cover the hill-slopes. All of this might be han-
dled, transported, marketed, to greater advantage if
right machinery were introduced. We have spoken
of stump pullers, such as are used in the United
States for similar work. As the custom now is to ship
not only the trunks but the roots of the logwood, the
gain of grubbing in a more systematic, nineteenth
century way would be enormous. Or, supposing the

root to be out of the ground, proper appliances for
crushing it would reduce its bulk, and practically
increase the carrying capacity of every car engaged in
shipping it by just so much. Or, better still, by the
investment of capital in a properly arranged plant for

extracting the dye from the logwood *before* it is shipped, the industry would without question receive an enormous impetus. What we wish to point out is that here is an almost unworked field for the investor, whose predilection for agriculture, fruit growing or manufacture could be indulged in the security which a well-equipped, firmly-established railroad affords.

After passing Spanish Town and May Pen, with its fine iron bridge and view of the dry bed of a river that has found a subterranean channel, we soon strike

OLD MANOR HOUSE AT EWARTON.

a perceptible up grade and gradually rise to higher levels, and towards Porus, the present terminus of one section of the road.

There are now two main sections, the junction being at Spanish Town.   At the end of one is Porus, at the other Ewarton.   Both of these towns are in the county of Middlesex, the first in the parish of Manchester and the other in St. Catherine.   Beyond Porus it is proposed to extend the line, and in fact, the addition is now being built through the delightful hill country north of Mandeville, past Shooters' Mountain, and toward the region of Cornwall County, known as the Cockpit country, from its deep sink holes and wild, cavernous character.   The lines run by Balaclava, and across the Black River.   Here some of the most difficult engineering work of the road is being accomplished.   Through Vauxhall and Ipswich the road will extend to the Great River, and along that stream to the neighborhood of the Lethe estate and thence to Montego Bay, thus opening up one of the pleasantest as well as one of the richest parts of the island.

The other section, that which now terminates at Ewarton, will branch from Bog Walk, cross the mountains east of St. Thomas-in-the-vale, and follow the course of the famed Rio d'Oro, where the Spaniards were supposed to have had the secret of the Indian gold mine ; a tradition not unlikely to be well founded, as the region gives fair indication of gold. From the Rio d'Oro, the railway will run through the lower part of St. Mary's to the Flint River, continue to follow that stream to its junction with the Wag Water, and following that, reach the coast at Annotto Bay, going thence eastward up the coast to

Port Antonio. The projection of a third line to cross Trelawney and St. James, is also spoken of.

It was in the year 1843 that the Jamaica Railway Company was incorporated. The line was opened for traffic in November of that year. It was at first only operated as far as "The Angels," near Spanish Town, a distance of fourteen miles from Kingston, at a cost of £222,250. It had but one track. From then to 1867 work was virtually at a standstill. Then an extension from Spanish Town to old Harbour Market was carried through at a cost of £60,000, being opened to the public in July, 1869.

Since 1867 the history of the road may not be uninteresting, showing as it does the gradual reception of a new idea, the final appreciation of business energy, and the success which attends intelligent management.

After the opening of the extension the business of the company increased gradually, till in 1875, its revenue reached the sum of £24,200, a gain of £13,-478 in six years. In '77, Sir Anthony Musgrave, as stated in a previous chapter, interested himself in the affairs of the railway, entered into negotiations with the company and effected the purchase of the road by the government. At this time the capital represented was £267,250.

The permanent way was relaid and ballasted, water ways and conduits were opened to drain those parts of the road which were apt to be submerged, twenty-eight bridges were built and general improvements

carried on all along the line. What with stations, walls, culverts and numerous other additions and repairs the cost amounted to £107,260. The cost, including purchase, was £201,192.

The first year's work after the government purchase returned a net profit of £5,621. The railway continued after this to more than pay its expenses.

The next move was the extension to Porus, in Manchester, the present (1890) terminus. On the 2d of May, 1885, the extension was open to traffic at a cost of £280,924, loans being raised to provide the means for expenditure. It was said that the contractors, Messrs. Reid and McKay, made nothing by their enterprise, and an award of £13,731 was made to them by the arbitrators. The raising of a further loan of £70,000 was authorized by Law 16 of 1887 for proposed extensions, surveys and repairs. A loan of £26,000 was raised and an extension through Annotto Bay, from Bog Walk to Port Antonio, was reported upon in September, 1888. About the same time some American capitalists, headed by Mr. Frederick Wesson, made an offer for the road. It was sold to them for £100,000 cash, and £700,000 secured by second mortgage bonds on the security of the railway at four per cent. interest. The company pledged itself to extend the line as projected, and was empowered to issue bonds to the extent of £320,000, and to make further issues of £200;000 on the completion of each twenty-five miles of extension, till the full amount of £1,500,000 is reached. The transfer of the line was made in January 4, 1890, since which time, under the

general management of Mr. R. B. Campbell, the road
has continued to give satisfaction to the public as well
as to its owners.

# SANTIAGO DE LA VEGA.

FAMILIARITY with the history of Jamaica means intimacy with Spanish Town, which for over three centuries and a half was the capital city of the island. Seville, Melilla, Oristan, were one after another abandoned by the Spaniards, who soon founded this city, whose salubrious climate, pleasant surroundings and defensible position seemed to point to its site as the natural one for the seat of government. This was Columbus' dukedom; this the city of which D'Oyley was left in charge after the English conquest, the place of retirement for the aristocracy of Port Royal, the scene of social triumphs, of legislative strife, the site of Jamaica's finest buildings, now going into decay for want of occupancy and proper care.

Through the hospitality of the custos of St. Catherine's parish we were first introduced to the attractions of this quaint old capital, and in telling of its features of interest where can we better begin than with Rodney? Admiral Lord Rodney, the hero of 1781, the defender of Jamaica against the French under Count de Grasse, lives forever in marble under a cupola overlooking the main square of Spanish Town. John Bacon, who apparently had a colonial contract to fill the island with evidences of his art, bestowed upon

the statue of Rodney his ripest skill. For a century it
has kept watch and ward over the affairs of Spanish
Town till it grew to have a more than educational
significance. People spoke of it as a person, and
regarded it as a tutelary deity. More than all, they
had affection for it.

Judge, then, what the feeling of Spanish Town
must have been when Rodney was removed to
Kingston, and set up in the market-place there with
his face to the sea. There was mourning. Houses were
hung with black. A mock funeral was attended by
numbers of people, and a coffin containing the effigy of
the lost admiral was placed in the empty cupola. Seriously, the authorities feared a riot. They had taken
away the government ; they had destroyed the *prestige*
of the place; they had robbed it of its business ; and
now they had multiplied injury and insult by carrying
off Rodney. There, Spanish Town drew the line. It
refused to be parted from its idol. There is something delicious in this robust epic; the rape of the
statue, the protest, struggle, restoration and triumph
which followed. Now, the admiral stands once more
on his own pedestal, in a costume borrowed from
Grecian mythology, and admirably adapted to a warm
climate, his captured cannon at his feet, and the
garden-centred square of Spanish Town under his
eagle eye.

And this is what he sees. A great empty residence,
whose wide doors no longer admit the state, beauty
and fashion of a capital; whose great, empty ball-
room, where princes of the blood have been enter-

tained, is empty ; whose banqueting hall is melancholy. There are the House of Assembly rooms devoted to baser uses ; the court-house and the various offices all pleading their loss of importance. Do not think from

this that Santiago de la Vega is desolate, though abandoned by the government for less lordly accommodations and business conveniences in Kingston. We visited the churches and chapels, and found that they were attended by good congregations. We took our places with the worshippers at the Cathedral Church, which stands on the foundations of the old Spanish Red Cross Church of St. Peter, and if the mural decorations, crests and monuments of the interior of that interesting place shared the attention which should have been given solely to the prayer book and the sermon, let art and antiquity divide the blame. Here are buried former governors and great men. The arms of Effingham, and the sweet features of the Countess of Elgin were texts in themselves.

Spanish Town possesses a good almshouse, hospital, market, record office and constabulary depot. Its streets are well paved and clean ; its houses attractive, though not different from those of most other West India towns. Its population numbers 5,689. It is situated six miles from the harbor and thirteen miles from Kingston, on the banks of the Rio Cobre, a beautiful stream of considerable volume, the water from which supplies the irrigation canal which enriches a great many hundred acres in St. Catherine's.

There are in Spanish Town several lodging-houses ; and lately a hotel company has undertaken the work of providing lodging and entertainment at the Hotel Rio Cobre, just outside of the town. This house will accommodate twenty-five to fifty guests, and aims to give Creole comfort and good fare, with American management. The proximity of the beautiful Rio Cobre is a guarantee that the sojourner will neither lack for active sport or poetic pabulum.

We enjoyed several drives in the good roads about Spanish Town. One of these led us to the Rio Cobre dam and beginning of the irrigation canal. The dam is a long, depressed slant, over which the water flows with a continuous ripple. The flood of the canal finds its way by pleasant banks, under picturesque bridges and beneath long, even rows of over-bending cocoanut trees to smaller channels, until, at last, its ramifications reach through grazing pens, fruit cultivations and sugar estates, fertilizing and enriching all that section of country.

A little way beyond the dam we were conducted through a field, where a group of colored people vainly tried to dissuade us from advancing, by stories of ticks in the grass and impassable swamps. Beyond this we saw traces of an ancient avenue of tall trees, and the evidences of something more than negro occupancy. Then we came upon what seemed to be a ruin, but so overgrown with trees and underbrush that it was necessary to send a man ahead with a machete before we

could advance. The ruin is said to be that of the
house of the last Spanish governor of the island, who
fled from here when the island was taken by Penn
and Venables in Great Cromwell's time. It may
easily be true. There is every evidence of antiquity
in the trees and the weather-eaten stones of the
house walls, now netted by the interlacing roots of a
century-old fig-tree that overrides them. Who knows
what treasure may underlie the rubbish that chokes
the doorways and cumbers the wall gaps?

Another pleasant drive from Spanish Town is the one
which skirts the salt ponds and goes out to Port Henderson. On the way thither we have a fine view of
the Healthshire hills on the right, till that is shut off
by the nearer Salt Pond hill. Port Henderson is the
natural landing-place for Spanish Town. It is nearly
opposite Port Royal, and diagonally across from
Kingston. There is a fine view of the harbor and the
hills beyond. The Constant Spring Hotel, the barracks at New Castle, the houses that dot the mountain
side here and there, and, still further away, the cloudy
summit of the Blue Mountain peak, all can be enjoyed
from Port Henderson. There are very few houses
here, but a small garrison is quartered on the spot and
visitors often stop to enjoy the mineral bath, a natural
rocky spring housed over. The neighborhood is full
of deep caves. Into one or two we went a little way,
but others are so deep that they give back no echo
to tell what becomes of the stone thrown into the
depths. In one of these caverns tradition says that
Morgan the buccaneer concealed both men and booty

at a time when it was not convenient to have either exposed too freely for the public view.

Near Port Henderson is the Battery of the Apostles, named evidently by some one who did not know how many apostles there were; or perhaps, one of these apostles may have burst at some time. In late years the defences of Apostles' Battery have been materially strengthened, and magazines and armament added. Four miles from Port Henderson are the defences of Passage fort, the landing-place of the English conquerors, at the mouth of the Rio Cobre. Port Royal on one side of the harbor entrance and the Apostles' Battery on the other, are, however, considered the true and important defences to both the old and the present capital. At Green Bay is the quarantine station. Back of the Apostles is a lookout known as "Rodney's," where the great admiral used to watch for the French. Near here is also the tomb of Lewis Galdy, who was swallowed by the great Port Royal earthquake and cast up again alive. He must have been indeed an unpalatable morsel if both the earth and the sea rejected him.

# HALF-WAY TREE AND CONSTANT SPRING.

ABOUT a mile and a half from Kingston, northeast of the race-course, are the "Up-park-camp" barracks. Here are the headquarters of the West Indian regiment stationed, with the brigade and other military offices. The place contains good barracks, parade ground, swimming bath, hospital, and everything to make camp life as endurable as possible. The location is health giving and cool and the views fine. The place is well worth a visit.

From Kingston, past the race-course and exhibition buildings, the main road, upon which the tram cars run, leads out to Half-way Tree. This highway is dotted with residences, many of them occupying the site of former pens, the names of which they still retain. Here are trim white dwellings, with jalousied verandas, high porches, and chimneyless roofs, that slant on four sides instead of being built with vertical ends, like the ordinary gable roof of the North. Around the houses grow broad leaved century plants, segregated branches of palms, great blazing masses of scarlet or yellow bloom on flowering shrubs and trees, clusters of deep-hued mango foliage and groups of tree ferns or beds of glowing blossoms. The only visible drawback to these residences is the cloud of dust that is apt to roll in from the road. The Half-way Tree road is well kept and much travelled. It is the beginning of the great highway that crosses the island, passing Constant Spring, crossing Stony Hill, traversing the mountains of the interior, following the Wag Water past Castleton Gardens and joining the coast road at Annotto Bay.

At any part it is well travelled, but on the lower end of the road, especially on Tuesdays and Saturdays, which are market days, the passing peasantry become a multitude, a tide that flows in the morning and ebbs again at evening.

The women are numerically much better represented in this throng than the men. Women are the workers among the blacks in the neighborhood of Kingston. They carry the coal on the wharves, load and unload vessels, drive donkeys and mules with produce, break stone on the road, carry stone and other building material for house builders, wash, bake, dig in the fields, and, in fine, perform four-fifths of the labor that one sees done. If the men work they keep the evidences out of sight with most unaccountable modesty. These women are generally strongly built, short and energetic. One or two garments, well girded up, suffice them for decency. Their tread is a firm long glide, with a great deal of motion as far up as the hips, a side to side swing below the waist and above

GOVERNOR BLAKE AND FAMILY AT CRAIGTON.

that absolute steadiness and rigidity, so that the heavy
burden on the head is not disturbed nor endangered.
They will chase each other, quarrel, jump aside for a
passing horse or vehicle and never upset or spill these
burdens.   They are a merry lot, with as little knowledge of physical pain or fatigue as strong young cattle.
The Jamaican male peasant may look morose, but the
women are always ready to break out into a smile.
Among this crowd are donkeys with panniers, mules
and horses laden with all manner of freight, people in
vehicles and on horseback going at as good a pace as
though they had a clear road.   The pedestrians
(always in the road, as there is no pretence of a side-
walk) scatter, make way, close in again ; all rapidly
moving in one direction, keeping the same swift,
steady gait that some of them have used for twelve or
fifteen miles over the mountains.   One of them will
be glad to get a shilling for a load heavy enough for a
donkey, which she has " toted " ten miles to market.

Meanwhile the dust envelopes everything ; the grass
by the wayside and the leaves of the mangoes and
cotton trees are gray with it.   It follows in an eddy
behind the street car that trundles by, drawn by its
team of galloping mules, and settles on the dusky
throng that set it in motion.

Some day, somebody should write, if not an epic, at
least an ode, to the mango vender.   She it is who supplies the great staple of diet, and when the stock in
trade is becoming unsalable she prevents waste by
eating it herself.   Seated by the roadside she devours
mangoes ; walking along the road she eats them ; talk-

ing, bargaining, scolding, laughing—the ever ready mango fills every pause.

At the village of Half-way Tree, three miles from Kingston, there is a cluster of houses on a cross road; a church and burying ground, old and quaint, and a few shops. At a little distance are several fine residences and among them the King's House, or Gubernatorial mansion, now occupied by his Excellency, Sir Henry Arthur Blake, K. C. M. G. This is a large and attractive dwelling built after the style of the country, with upper and lower verandas shaded by jalousies, entirely enclosing it. The lawn is beautifully adorned with shrubbery, and handsome trees shade it. To Lady Blake's rare taste is due much of the floral wealth that charms the visitor to King's House. There are also, in the neighborhood, the homes of several of the higher government officials. Near the cross road

is situated the market building, and there are also a court-house or girls' orphanage asylum and school near by. The church referred to above is one which was built in the reign of Queen Anne, and has been in recent years repaired and renovated. Its floor is paved with slabs inscribed with the names and crests of many departed worthies. Sir Nicholas Lawes, once governor of the island, is buried there. In this church, Livingstone, signer of the Declaration of Independence, is said to have been married, and in the churchyard reposes all that is mortal of Colonel Harrison, grand uncle of the present (1889) President of the United States. Colonel Harrison represented the United States Government in Jamaica for a number of years. Upon the 30th of May last, a few Americans, headed by Mr. Estes, their consul, engaged in a simple decoration day service over his grave.

The tram cars, which run from Kingston to Half-way Tree with a headway of twenty minutes, start from the latter place to Constant Spring every half hour. This is another three-mile drive along a road which becomes cooler and pleasanter at every step. The open roadway ahead and the glimpses through gaps in the foliage reveal more and more of the mountains,—those mountains of which Columbus' crumpled bit of paper was at once a description and a parody. Ravines that are mere lines near the summits and widen into valleys as they approach the plain, are the recipients for streams from other ravines. Add to this effect the winding, irregular shape of a range of hills and the whole has a singular crinkled appearance. But the warm, rich tones that sweep across their under coloring of blues and grays, combined with these convolute forms, give effects that are often startlingly beautiful to the Jamaica Mountains.

The other road from Half-way Tree, going to the east and north, leads to Gordon Town, about six miles distant. There is a daily stage from Kingston to Gordon Town. It is a small village, having little of note about it. But from there one may proceed on horseback (a good livery establishment is there) to Newcastle, well up on a Blue Mountain spur, where the white troops are stationed. This is 3,974 feet above the sea. Between Gordon Town and Newcastle is Craigton, the mountain home of Governor and Lady Blake. In this vicinity are also mineral springs at a place called Silver Spring. There are two or three points of interest on the road to Gordon Town. Near this place is quite a large coolie settlement, where these gentle East Indian laborers live in something of the fashion of their eastern home. Here the habits of India (at least of its laboring castes) may be studied as well as on the banks of the Ganges or the Indus. They keep the same distinctions in costume for age or rank that obtain in Calcutta. They are, many of them, skilful artisans, and the visitor may see the necklace or other ornament fashioned from the handful of silver pieces he furnishes for the purpose "while he waits." Intellectually we need not say where the coolie stands in comparison to the negro. He belongs to a darker branch of the Caucasian race, his civilization one of the oldest in the world, and though of a lower caste in that race yet far above the African in development. It is amusing to notice that the negro looks down on the coolie as upon an inferior.

Between Gordon Town and Half-way Tree are the Hope Gardens, which will be elsewhere described, and the Hope high school for boys, a large and somewhat imposing building, erected on the government ground. Near here are also the lead works and reservoirs of the Kingston and Liguanea water works.

These are the principal roads and villages in the parish of St. Andrews. At Constant Spring, which is the terminus of the tram car line, there are the water works, built when the Wag was taken to supplement the Hope supply in providing Kingston with water. These works are called the Kingston and Liguanea water works. Constant Spring also boasts a great hotel, the largest building of its class that Jamaica has ever known. Its centre and two wings contain ample accommodation for several hundred guests. Its cuisine is presided over by a competent French chef, its tables are well supplied and well served, and the whole place is under excellent (American) management. It is but just to say that this hotel was the first building in Jamaica where electric lights were used. Its charges compare favorably with those of other hotels and lodging-houses in the island. We have elsewhere given the ordinary rates of board and lodging.

# STONY HILL AND CASTLETON.

IT is a rare drive for a May morning from Constant Spring to Castleton. The heat and glare and dust, which annoy the traveller on the half-way tree section of the road, disappear entirely when we have passed Constant Spring. The air is fresher, the fields and foliage greener, the light pleasanter than on the lower ground. We have reached the limits of the plain, and before us rise the crumpled, irregular hills that slope back towards the Wag Water, or rise, fold and convolute fold on ridge and spur, till far in the distance they reach the highest island altitude of over seven thousand feet, in the Blue Mountain peak.

If there is anything calculated to make a man glad he is alive, it is to mount a good horse on such a morning as Jamaica knows in May, before "the rains" have made up their minds to restore the diluvian epoch, and commence the ascent of that picturesque pile of geology and verdure known as Stony Hill. There are a few grazing cattle in the fields, a house or two by the wayside, a side path that invites one—with that seductive way that shady side paths have—to turn astray from his road.   After that a sharp gallop over a tolerably level piece of good road brings one to the Liguanea water-works, at the foot of the hill. Here an well-fashioned stone aqueduct conducts the water of the " Wag " into a reservoir, from which the distant city is partly supplied with the fresh, pure water, which is one of its chiefest blessings.

From this picturesque structure the way leads by small banana cultivations, and under the shade of the ubiquitous mango trees, where the temperature is delightfully cool and fresh, and the light very restful to the eye, as it is transmitted through the broad, bright leaves of the banana plants.   Still rising, we reach a narrow place, where, by engineering skill, the good road has been cut and walled up to the mountain side, and protected by a substantial rail, so that the unwary rider may not find his descent of Stony Hill more rapid than comfort actually demands.

From this turn the first extensive view greets us. The mountains of the lower range, or Ramshorn Ridge, stretch away to the right.   In the middle of the picture is a single hill, that wraps its drapery of verdure about it in solitary dignity; between these two are the upper end of the Liguanea Plain and dis-tant Mona Vale, with its fresh and abundant green marking the limit of the sugar-cane fields ; to the right lies the greater part of the plain and the sea. There, eight miles away is Kingston ; further still, the line of the palisades, the light-house and the vessels, that look like floating specks on the blue expanse of the distant ocean. Further up Stony Hill, we find, at a

sharp turn, teams to pass (which we do by following the law of the land, and turning to the left), and sudden glimpses of the valley, seen through, or over, the tops of orchid-ridden trees.

Near the top of Stony Hill is a little settlement, a few cabins and stores, and here a road to the left leads to the grounds of the reformatory, large buildings originally used as barracks, but now the home of a little army of bad boys, who wear their uniform of disgrace with nonchalant composure, and seem to be altogether better cared for, and better fed, and cleaner than boys of the same class who have not yet been put under the government's fostering care. The houses of the reformatory are large and airy, built of white stone, and commanding a fine view of the plains and coast below. The superintendent denies that his boys are bad boys—" only a little wild and unruly," he says.

The mists are still clinging to the distant ravines and glens, as we ride along the ridge overlooking the vale where the Wag Water flows. And what a sight it is! Here are cane fields; there, acres of tobacco; again, gardens full of all the fruit and vegetable produce dear to the native Creole palate. Groves of cocoanut; miles of plantain and banana; hillsides covered with ferns; houses, part wattled and part daubed with colored clay; red flowers of the orchid, glowing like spots of flame from the cottonwood branches; women striding along under their burdens, destined for the market in Kingston; donkeys, carts—everything, in fine, that can occur to one as being desirable to enliven or beautify a tropical landscape on a May morning.

At several points are sharp turns, where the road follows the dip of some almost vertical ravines. There is an ancient and massive look about the safeguard walls of these places; but the most picturesque, quaint and solid looking piece of engineering on the road is where a well buttressed bridge spans the Wag Water lock at the head of the water-work system. From it the view of red roofed houses, nestled among the living green, the flashing water, the never ceasing variety of luxuriant bloom on the hillsides and by the river banks, combine to satisfy the sense, and still to excite the imagination.

" It is a far call " to Castleton, but the way seems short by a road where the spaces from milestone to milestone are packed with interest and paved with beauty.

And what is Castleton? Far in the interior of the island, occupying a central position between the north and the south coast, on the road that runs from King-ston to Annotto Bay, nineteen miles from the former place, nature made a garden with all the advantages of loveliness and fertility that a rich valley and a beautiful stream could combine to furnish. Its slopes were blessed with a sufficient and even rainfall (109.35 inches annually); its soil was rich and deep, its cli-mate never cold, nor ever uncomfortably hot, the mean temperature being 74.5° Far. That people named this spot of ground Castleton and not Eden, proved them inapt at drawing parallels.

On such a basis of natural advantages and beauty the government built, wisely, a fairy garden, a sort of tropical Kew, where all the strange and useful plants of other warm countries might be fostered and made to feel at home. Here bloom myriads of native and imported orchids. India and the isles of the sea have been called upon to contribute their valued foliage, and food plants, and medicinal trees and herbs. There is a large industrial ground for novel economic plants.

The nurseries contain about 40,000 plants, such as cacao, olive, sugar-cane, rubber plants, nutmeg, clove, black pepper, mango, vanilla, cardamon, pineapple, sarsaparilla, cinnamon, Liberian coffee, tea, etc. Taste and skill have combined to arrange these beautiful trophies in a manner pleasing to the eye, and good sense has dominated the arrangement so that the visitor may feel at his ease and find comfort on the benches that are placed along the well-kept pathways, or enjoy from the shade of its vine-covered arbors, the swift Wag, as it loiters in rich circled pools, or leaps by in eddying rapids.

Castleton is a good place to visit, either for purposes of social recreation, or more serious labor. The landscapist, the botanist or the mere picnicist are alike satisfied here.

# IN THE HILLS.

"How far is it to Cherry Garden?"

"Na *too* faa, maastah."

"Thank you. Where does this road go?"

"'E go wehevah you wansteh go, sah."

"Accommodating road."

"Yessah, anywhere you wishes—"

"Does it go to Hope Garden or Constant Spring or Mona?"

"No sah, none o' dem places, it jis go wehevah you wants to go an' 'tain' too faa."

One fork of it led to a gully; the dry, rock-bedded, tree-bordered channel through which the vernal and autumnal torrents from the mountains find their way to the sea.

Growing wild on every hand you will find the prized hot-house treasures of your northern home. The color box of the artist cannot outdo the hues and shades of the scores of species of orchids. Lantanas, blue and purple, yellow and red, some dwarfed, others gigantic, are scattered with the lavish hand of a gardener who fears no frost nor dreads a drought. Wild peas, of colors galore, and with a tropic vigor that fails at no height, bedeck the fallen sugar mill or trail along the cactus hedge. Species of sensitive plant, called "shame" by the pickaninnies, and their first cousins, the giant Cassias, are features of every meadow path or distant vista. Here the locust tribe present never ending surprises for the student of the few modest species in our northern fields; truly they seem to vie with the orchids in the variety of form and color of their flowers.

Oxalis, too, and Ranunculus are occasional reminders of that land where winter blasts and angry winds have at last driven to these wilds of everlasting June.

Then is no word, nor any combination of words, that will tell how Dame Nature has enriched these her footpaths in the everlasting hills.

Wherever the sheer mountains have set a brook as a boundary between them, straightway there has flowed a full stream of exquisite, multitudinous life; a fountain of youth, where we find our boyhood waiting for us with the old holiday zest of discovery and adventure.

Let us warn you, comrade, that like the country that honest John Bunyan saw in his vision, this domain has castles dangerous and many a hill of difficulty.

Glens and jungles in these mountain slopes are only open to the most intrepid and persevering lover of nature. The Spanish dagger, with its needle-tipped leaves and recurved sping serrations, the many species of prickly pears and other cacti, all abounding in bristly or pilose coverings, with the most excoriat-

ing powers; the dreaded "cow-itch," a powerful
creeper with a pod whose bursting fills the air with
myriads of floating spinules, penetrating and poisonous
to the extreme; the various heavy-trailing vines,
parasitic and parasite bearing, catching the foot at
every incautious step; the hidden ants' nest beneath
the crumbling loam, undisturbed save at nature's great
plowing time, the season of earthquakes; or the still
more to be dreaded ant citadel in the crotch of the
sapling to which you cling for temporary support:—
all these and many other agents unsuspected torment
and combine to make the unwary and unobserving for
ever after a closer student of the wonders if not of
the beauties of nature.

On the hither border of this domain is an ancient
stone aqueduct, terminating in one of the most picturesque ruins imaginable, the remains of an old sugar
mill. The gay foliage of the trees and creepers contrast brilliantly with the cool gray tones of its walls.
The arch of its upper windows frame mosaics of blue
sky and vine leaves. Nothing is left of the wheel
now but the wreck of its axle, black and mossy.
Naught to show where the water flowed from its
sluice but the massed stalactites at the further end of
the well.

A woman near by is grubbing yams in a field, and a
man with load on head and machete in hand stops
to speak to her. No living being, born outside of
Jamaica, could possibly understand their *patois* or
imagine it to be a dialect of the English tongue.
Their voices alone break the silence where, a hundred

years ago, there was the constant hum of industry; for this was the centre of one of the great sugar estates whose miles of cane fields waved where now the occasional hut of the free negro barely breaks the thicket. Can we, in fancy, re-establish the scene? Can we not imagine the drone of the wheel, the song of the workers, the soft plashing of the stream, and, above all, the fragrance of the bruised cane or the river of amber sweetness fed by its rivulets of sap?

Drawn by these memories of sweetness, flits and hovers that most fitful of all butterflies, the Banded Heliconia—*Heliconia Charitonia*, of the savants. Who can lead a life freer from care than does this prince of idleness?

In these wildernesses the ants abound in numbers far beyond the apparent needs of Dame Nature, and sufficiently to afford a constant menace to him who would cast himself into her verdant lap. Ants that are black; and yellow, and red, ants that crawl and ants that fly; ants that work and bite and sting—all and others are here. Tunnels and turrets and towers attest their architectural skill. Great paper and mud nests, high in the trees above the floods, speak for their foresightedness, and certain of their number herding their "cows," species of plant lice and of butterfly larvæ, bear witness to their thrift and ability as masters.

Nature is not directly responsible for the presence of all the ants here in Jamaica. Some have been introduced in the wealth of vegetation from other lands, for which this paradise of the botanist is

famous. Formica Omnivora, at times terrible in its
depredations, as its name well indicates, was introduced here for the purpose of ridding the island of
other insect pests. As in the case of the English
sparrow and the caterpillars in our Eastern States, it
was introduced to do just what it was least likely it
would do, and, still like its ornithological prototype, it
has become an ineradicable nuisance.

Leaving the gorge and skirting a field of pineapples, we begin the ascent of an almost perpendicular
hillside. So steep is it, in fact, that it is impossible to
make it in a straight line. We must advance by
zigzags towards the summit. Half way up the hill—
where we do not arrive until after several breathing
rests—there is a wattled hut roofed with dry banana
leaves. Still above this we strike a patch of guinea-
grass and hills of yams ; that is, if it is proper to speak
of moles and corrugations on the side of a vertical
plane as "hills." The mystery is how any one ever
succeeds in teaching yams to grow in that position, or
how the soil manages to adhere to the hill frame without being nailed fast. Nor does it always so adhere.
Many a truck farmer—save the mark !—has awakened
to find that the trembling of the earth which disturbed his slumbers was his yam field *in transitu*
towards the clearing of his neighbor in the valley
below him.

Look across the valley. The well-known engraving
by Hogarth, in which that artist turns to ridicule his
professional brethren's proficiency in perspective, is
outdone by the prospect before us. There, on the

opposite mountain, hardly more than a pistol shot away are fields and gardens like those we are traversing. They are such as appear in all the slopes; spots and patches where the insect, Homo Africanus, has

scratched the natural verdure with his hoe, till, at a distance, the upper country seems to be affected with a mange. But the singular feature of this industrial display is, that almost without exception these gar-

dens are wider at the top than at the bottom, the
result being a total inversion of usual perspective
effects. There is nothing to which we can liken it,
but the impression (which we all learned in boyhood
to appreciate) of a landscape seen when one is hanging
head downwards from a horizontal bar.

In these upper reaches of the foot-hills we are con-
fronted on every hand with the testimony of the rocks
to the upheavals of nature, that were necessary before
these crenated and water-washed hills were raised
from the bosom of the troubled waters, from beneath
which these portions of the lost Atlantic have at last
been rescued.

These, too, are the mountains of lost springs and
rivers. Here the mountain torrent, white with the
foam of boundless energy beneath a southern sun, and
tired of the beating ray and blistering rocks, dips
down into the cool cavern and is forever lost to sight.

From the bridle road that girdles the hillside, we
drop again by a tortuous, slippery path to the valley.
So precipitous is this descent that it seems foolhardy
for any one but a goat to attempt it. Yet on this
trail, and on a thousand like it, sturdy, barefooted
women, carrying heavy burdens of fruit, pass and
repass daily. Presently we come to a cabin, built
mostly of bamboo and palmetto. Its sides are woven
like a basket and its roof is a thick mat of thatch, the
whole structure being apparently glued to the hillside,
which has an angle of about seventy-five degrees. In
front of the hut stands its owner, a jolly-looking
black fellow, who is engaged in an animated conversa-

tion with a young woman across the way. And we wish to explain, parenthetically, that " over the way " means just across the valley on the opposite mountain slope, to reach which one must accomplish about three-quarters of a mile of up and down hill walking. He is saying: " What you is doin' yonner topside ? " Topside is very expressive, but we always supposed it to belong to the " pigeon " vocabulary of John China-man.

Her answer is unintelligible ; not because of the distance, but for the reason that we of the sketch book and the insect net have not yet completed our education in unconsonated English.

He responds : " Ah ! you bin tief banana." Which is evidently a pleasantry, for he turns to us with a merry smile on his glistening, ebony countenance.

" How you is dis maanin', maastah ? "

Getting the needed information from this cliff-dweller, we follow a path that is bordered on both sides by banana plants on which the green fruit is hanging. This is what is here called a " banana walk." Each plant bears a single bunch of fruit placed at the base of the leaves ; from it depends a heavy plummet-shaped growth about the size of a large orange, and reminding one of an overgrown and undeveloped rosebud. Its color is a deep maroon, and a detached petal or two prove it to be a blossom.

Luxuriant nature encourages idleness ; idleness is but a step towards parasitism. Here, in these wild tangles of vine and mighty growths of tree and under-brush and fern, parasites of the vegetable world

abound in every direction. From the mighty silk-
cotton tree to the tiniest fern, all seem at times to be
subject to the sapping and deadening effects of some
parasitic growth. Among the commonest of these are
the many orchidaceous plants, whose varying flowers
of endless form and color vie with those of the gaudy
butterflies and moths that are attracted by them.
Jamaica has furnished a number of orchids new to the
botanical world, and it is deep in these haunts that
these rarities should be sought. No words of ours
can picture the charm of many a sturdy trunk or limb
of some woodland giant as it towers aloft, draped in
these daintiest of vestments. Surely no family of
plants presents such variety of form and color in its
blossoms as do the orchidaceæ; some there are that
so closely mimic the more brilliant butterflies as to be
readily mistaken for them.

The "air pines" are another tribe of plants that add
largely to the picturesqueness of these filmy haunts.
"Pine," you must know, here in the Antilles, means

not a tree or even a bush, but applies only to the pineapple tribe, and those plants which mimic them in foliation. Wild and cultivated pines there are, bearing apples of varying flavor and juiciness, from the "bastard" to the Ripley pine. Where the line of demarcation between a state of unkempt nature and cultivation lies, it will always puzzle the uninitiated to discover. Frequently cultivation seems to signify naught, save a very slight degree of oversight to prevent "buckra" from helping himself to the fruits which he supposes to be growing in an uncared for wilderness, surrounded as they are, by every form of tropic vegetation which will grow in the dryer soils of the slopes.

Down a wood-path, bordering a never failing little stream that years ago gained for itself the name of the "Constant Spring," a name which has extended to the great sugar estate which once depended upon it, a ripe golden sided mango hangs temptingly overhead. The mangoes are the common food of the common people and the prized dessert of the better classes; and certainly nature has been generous about the supply, for you cannot go many rods in any direction without finding mango trees growing, either singly or in groups, and all laden with fruit which in varying flavors will last from late March to Christmas. But between the delicious "number eleven" and the commonest sort, there is as wide a difference as between a gnarled, no-account little cooking pear and an old fashioned thoroughbred Bartlett. The first is nectar, the second turpentine.

The mango hanging over the way does not hang
there long. It is not a number eleven; but it is
delicious. So much so, that it encourages a taste for
further fruit stealing. That is the best of this country; it is the paradise of the fruit lover—the land
where that rare old alchemist, the sun, packs earth's
most delicate and fragrant essences in most attractive
shapes.

One of our friends spends his time in impressing
upon people the fact that here the banana and the
pineapple are side by side, and the mango and the
naseberry lie down together, as it were, while the
sugar cane and the sapodillo are inseparable, and all
can be had for the picking. It does not seem as
though living ought to be expensive where most of the
people never saw a store, where many of them never
wear more clothing than the law absolutely demands,
and where the most desirable food is so convenient
that all the native needs to do in the way of work is to
sleep under a tree with his mouth open.

Only at night, however, can the romantic seeker
after Faiy-folk (the "filmies" of Jamaica folk lore) hope
to see nature in a mood when they do most unbend
and elfin pranks and brownie gambols are the order of
the hour.

Follow yon giant briefly; one of those monstrous·
fellows whose single beam lights up the student's page,
or sheds a halo around the dusk beauty's head, lashed
as he is in a harness of ·horsehair to the curly locks of
some belle of the ballroom. As this one—our guide
let him be—flits hither and yon in search of his less

brilliant mate, a ray of his light gives us a momentary peep at a fairy rendezvous; and just as it fades away we have discerned a score of tiny forms in gala array, marshalled near a clump of filmy ferns. Were they the filmies at last, or were they some insect host seen by the flickering light and through the eyes of a nature-lover's imagination? Alas! the fitful gleam that revealed them to us has made known our presence to them, for by the next ray from this entomological flash-light we see but the empty sod where but a moment before were those mystic forms.

Just as we resolve to step with gentler tread, a great, brown owl, that night-watchman who never needs the aid of burglar-alarm or time-detector, emits his all pervading cry, "Oh! ho! Oh! ho! a man below!" and then a fluttering sound—had those filmies wings?—and a sound of scurrying and scampering tells us that the spell is broken and we of all the human-shaped world are alone in one of nature's weirdest habitations.

Jamaica, possessed of 500 miles of coast line in which bays and harbors abound, with but 646 square miles of plain and about 3,550 square miles of mountain land is an extreme illustration of the varied topography which depends on volcanic origin followed by centuries of erosion. Here the landslip, the endless chain of caverns, the sinking river, the resistless mountain torrent—raging to-day and dry in a few hours—all add to the difficulties which tropical vegetation and unsolved meteorological and geological problems have already brought to the student.

The island contains but 4,193 square miles, these being less than one sixty-fifth the size of the State of Texas and less than one-third the size of San Diego County, California. Yet, small as is its area, so varied is its contour and surface that it presents a most diversified range of scenery and climate.

To treat of the Geology of Jamaica in so narrow limits as are here at our command is quite out of the question. The reader who desires to obtain an intimate knowledge of this feature of the island, will do well to consult Sawkin's "Geology of Jamaica," which, though published some years ago, is still the standard authority on this subject. Suffice it here to say that the basis of the island is igneous in character, deposed upon which are several distinct formations. White and yellow limestones, carbonaceous shales, trappean and metamorphosed series, conglomerates, porphoritic and granitic rocks, alluvial deposits, with numerous traces of iron, copper, lead, manganese and cobalt and some signs of good marbles are so interwoven and generally distributed as to make the effect on the student quite confusing at first. The surface of Jamaica, as is attested by the but 646 square miles of flat lands, is very mountainous, and in places quite rugged, though always verdure clad, save where an occasional washout or landslip has temporarily denuded the face of some precipitous hill. Running nearly due east and west, the island has a midrib or backbone of mountain peaks, the culminating point being reached at the peaks of the Blue Mountains, south of Hope Bay in Portland and on the dividing line between that parish and that of St. Thomas ye East. The highest of these three peaks reaches an altitude of 7,360 feet, which is somewhat higher than any other land in Eastern North America.

Several of the subordinate ranges approach the main chain both in height and scenic grandeur. Thus from Albion, in St. Thomas beyond Bellevue in St. Andrews, the St. Catharine's range presents a varied and rugged face towards the south, raising to its greatest elevation at Catharine's Peak north of New-castle—5,070 feet—and shutting out the higher peaks of the Blue Mountains from the plains of Liquanea and Rio Cobre. So, too, the plains to the north of Yallahs, in St. Thomas, are for the most part shut out from a view of the Blue Mountains by the Coward's and Queensbury Ridges, which, at Yallahs Hill, culminate in an elevation of 2,348 feet. So abrupt are these cross and parallel ridges, and so woven together in a network of peak and valley, ridge and ravine, that it has been impossible for us to more than indicate the position of a few of the principal mountain sum-mits.

Jamaica is much higher to the eastward, and greatly reduced in general elevation for a long stretch before the western limit is reached. This is interesting to the student of mountain ranges and island formations, as it is one of the very few exceptions to a very gen-eral rule to the contrary long ago pointed out by Dar-win and others.

The well timbered and vine clad hills are every-where riven by the streams, which become raging tor-

rents in time of heavy rains and which during droughts
are but empty cañons or gorges of great beauty and
wildness.

As elsewhere stated, these rivers are notable for the
remarkable fact that many of them never reach the
sea. As in the cave regions of Kentucky and Tennes-
see, so in Jamaica there are a number of rivers that
disappear into cavernous openings in the earth, some
never to appear again, so far as is yet known, others
only at considerable distances, and under new names.
Probably the most remarkable case of a lost stream is
that of Hector's River on the southern boundary of
the Parish of Trelawney, which after about twelve
miles on the surface, suddenly drops in the wild
honeycomb formation east of Accompong Town, only
to be resurrected at Oxford in Manchester under the
name of the One Mye River.

After another brief surface course it is lost to
sight under a ridge of the Bogue Hills. From
the other side of this ridge it emanates at both
Mexico and Island Estates in the Parish of St. Eliza.
beth, rushing forth with a considerably increased vol.
ume. From these points for a space of forty-four
miles now known as the Black River, it winds its way
through the Savannahs and swamps until it reaches the
coast at the Bay of Black River. This is unquestionably the largest river on the island, and it is the only
one on which navigation can be carried on at all times
of the year for any considerable distance. It is not
by disappearance into caverns that a number of the
rivers of Jamaica are lost to sight, but by actual evap-

oration, as they run over the pebbles and between the boulders under the scorching and blistering tropical sunlight. It is a common occurrence to find that some ever-flowing mountain stream, vigorous even in times of drought, will slowly dwindle away as it emerges from the shadows of the heavy woods of the mountain gorges. Such a stream, so easily evaporated in the dry season, reaches the coast after the terrific downpours of the wet season with resistless volume and velocity. Then it is that these usually very insignificant creeks or brooks earn their right to be called rivers. Fortunately the Department of Public Works is fast putting up good and durable bridges over nearly all the streams which are really formidable barriers to safe travel during the season of rains, and soon there will be but little of this sort for the traveller to dread.

Well watered as is the greatest part of Jamaica, still there are extensive districts in the middle and western parts of the island which are almost barren of water. This is not due to the entire absence of water courses, for there too the ravines, usually dry, carry off a considerable rainfall during short periods, though there are long stretches of weeks and even months when no rain falls, leaving the inhabitants in sore distress. As the principal mountain ranges trend east and west, it is plain that most of the rivers will take a northerly or southerly course. At the eastern end of the island the Plantain Garden River, and at the northwestern extremity the Montejo River are the only notable exceptions to this rule. A further discussion of the rivers of the island, with descriptions of their peculiarities and beauties, will be found in the general description of tours among the parishes.

The geological structure of Jamaica is such as to lead the observer at once to look for caves, and throughout the island, though principally west of the Parish of St. Andrews, numbers of these, many of which are in all probability connected by long series of subterranean passages, will be found.

In the early days of Maroon warfare they were constantly employed by that freedom-loving people for the purpose of eluding their enemies. It is also probable that many of those nearest the seacoast were used by the Spanish and Portuguese buccaneers as rendezvous and for the storing of their booty. It must be confessed that it is from this point of view that these caverns are looked upon with most interest by Jamaicans, and the negro fables and superstitions attached to some of them would furnish the romancer food for many a stirring tale.

In common with all cavernous countries Jamaica is almost devoid of lakes ; none really worthy the name are to be found. Ponds, both fresh and salt, there are , usually the latter, along the south coast. A lake has been recently reported as present near the summit of the John Crow Mountains, in the Parish of Portland, but judging from the description given by the intrepid explorer who visited this wild and unsettled region, it is rather to be set down as a large sink hole, or perhaps the eroded crater of a long extinct volcano. Sink holes there are without number, several of them being both by situation and formation very weird and picturesque. Many of these receive notice in other chapters.

# AN EARLY START.

STARTING from the American Hotel, at Constant Spring, early on a June morning (for this is a summer outing), we commence a long promised tour which shall take us through all the coast towns and villages of Jamaica. An inspection of the "buggy," or two seated covered trap, results satisfactorily. Richard Davis, the best driver in Jamaica, is on the front seat, and beside him is piled the necessary luggage for a three weeks jaunt. The horses, able and willing to do all the work required of them, wait for us without any of that unreasonable anxiety and impatience which less experienced horses might show. They have never been abused nor worn out, but they are not novices at long distance pulling. They will take us over mountains, around coast roads, through valleys, in sun and rain, for four hundred and fifty miles and be back in twenty-one days, very little thinner or less serviceable than now. Richard, too, deserves a word of special notice. A perfect guide-book and ready reference is he, furnished with marginal notes and bound in brown leather. Having driven "His 'Xc'llency de Gov'nah" he has acquired dignity; being intelligent and humane, he drives skilfully, without injury to his charges, the horse-kind; and as for the human-kind under his care, we gave ourselves up with a sigh of relief to his ministrations. We took no thought for the horses feed, nor the location of the best lodging-houses, nor the exact position of points of interest. Richard would not fail to furnish the desired information when required. He is the only boy of his class that we ever saw who can give an accurate estimate of the distance between two places.

A few hand-shakes and we were off. The sun had risen before us, as we were late in starting, as people usually are on such occasions. But those who imagine that a June morning in Jamaica, even after the sun has risen, is anything less than delightful, must have had his mind poisoned by false reports, or else be so puffed up with pride over his own particular climate, wherever it may be, that he has no just appreciation of any other sort of weather.

While we are jogging down through Paradise Town towards the Windward road, in the dewy freshness of the early day, it will be a good time to say a few words to the people who are forever pitying the man or woman who outstays the winter in this latitude. Of course the midday sun is hot. Of course people who are not used to it must expect headaches, and even fevers, if they exercise in it too much before becoming accustomed to it, or without taking the proper precautions as to clothing and drink. Of course it is hot enough to cook eggs. What would you have in latitude 17°?

But be assured that, having taken fewer precautions than the Northerner expects of his brother from the tropics, who visits him in midwinter, after a few weeks residence we have walked for ten or fifteen miles over the hills, through the hottest part of the day, with an unclouded sky overhead, *and enjoyed it.*

But this is sunshine we are talking about. Come into the shade, and you will laugh, as we did, over the discomfiture of New York, Philadelphia or Chicago people, who are sweltering in an atmosphere of humidity with the thermometer 'way up in the nineties, while yours registers 83° or 84°, and the constant breeze is dry and pure. There are some old notions that must be reconsidered, old judgments that will have to be reversed. That fallacy about Jamaica's climate is one of them.

We are swinging out between the pound and the lunatic asylum, past Chelsea pier, towards Long Mountain. There, high up on its side in the sunshine, is a bare rock that looks like a house, and, further on, a house that might be mistaken for a rock, with somewhere near them the trace of a road. As we near the harbor head we see a large barge, rowed by convicts, making its way for Rock fort. The landing and discharge of its living freight is a surprise. . Where under the ——deck—— did that long line of prisoners come from? They march, two and two, to a quarry, near which we pass. They are dressed in loose, cool-looking white sacking, and each one has painted upon the back of his upper garment certain cabalistic signs, figures and letters, which give more information regarding his name, rank in crime, and condition of servitude that we have pretended to remember.

Close to the quarry is Rock fort, an impotent but picturesque piece of military architecture that the Spaniards set much store-by, till they found that the shot from English guns found their way over it. It is not useful at present, though highly ornamental, and the arch at the rear, under which we on the road must pass, is fairly green with moss and romance. Rockfort is one of the most perfectly satisfactory things in its way that man ever imagined.—The view of the fort from the harbor is fine, and the view of the harbor from the fort beggars description. But we could not help wondering how those poor rascals bending over the stone piles, with their backs marked like so much merchandise, regarded it. The masonry is in good repair and the place is occupied by the colonial department. To the unmilitary man its reason for being is a mystery, since at present it guards nothing.

# ON THE WINDWARD ROAD.

HAVING passed Rock fort and the harbor head, skirted the foot of Long Mountain, where its magnificent wooded flank sweeps in a horseshoe curve from east to south, and forded the uncertain waters of the Hope River, we come to a wide, stony river bed, dry but for a miniature stream that steals along, under a fringe of bushes, to the ocean. Half a mile away, upon the other side of the stony level, commence the foot-hills, that rise rapidly and acquire dignity and proportions till they look down upon us from the cloud society they have got into.

We have left our carriage in the shadow of a tree near the roadside, and with an admonition to the driver to keep watch, start for the gorge, that must lie somewhere beyond that first wooded spur to the left. As we turn the point the view is striking and impressive. On the right the morning shadows still lie, while the mountain at our left is radiant with sunlight. The water worn rocks over which we are passing, tell their own story of sudden flood and overflow, and as we get into a narrower valley where it begins to close up to the ravine a mile or two beyond, there are further evidences apparent of strong floods. Here and there the earth is cut away, making a sharp terrace.

Then we approach closer to the rocks and see a well-worn water-mark, higher than a man's head. It seems hardly possible to believe that the little brook at our feet can swell to that. Further still, and we enter a cañon whose walls, several hundred feet in height, rise sheer on either side. There must have been a grand convulsion of nature here at some time, for on our left the strata of the rock may be seen almost vertical, clear to the top.

As we ascend, the stream grows larger. On the plain much of it has evaporated. We come to falls, piled up rocks, whirling-pools; looking into some of these we find that they are full of Mountain Mullet, a treat for a sportsman; then a narrow gateway, beyond which the gorge widens a little, and we mount by a well built stone pathway, a sort of Jacob's ladder, through a cave and under an overhanging rock to the principal fall. It is a spot to dream about—to be enthusiastic over; and yet we found few people in Kingston, less than ten miles away, that knew of its existence. Most of them could tell you more of some cascade in the highlands of Scotland, or the Catskills of New York, either of which might play second to this dainty bit of Jamaican scenery.

Occasionally a party drive out there to picnic. That is one good thing about Jamaica : a picnic party leaves no sign. The ubiquitous negro appropriates anything in the way of tin or paper, and John Crow takes care of all other fragments. Returning to our trap we are soon on the way to Yallahs, crossing by the way of the Yallahs River, a stream much like these we have described only larger than they ; it is too wide and too shifting to be successfully bridged, and so it is very likely that passengers will continue to take the chances of sudden floods and the danger attending them. Not long ago a whole family was lost on this spot, and frequently those travelling this road have been caught for two or three days between Hope and Yallahs, in a region where lodging houses are an unknown quantity.

Easington, the ancient capital of the parish of St. Davids, before that parish was merged into St. Thomas, lies inland on the Yallahs River. It has quite a fine suspension bridge and is reached by a very fair road. The court meets here twice during the month. It is one of the five principal towns of St. Thomas; the others being Bath, Port Morant, Morant Bay and Yallahs Bay.

Just before reaching Yallahs River we come to the great sugar estate of Albion ; broad acres of growing cane, through which at frequent intervals flow the irrigating streams, so necessary to successful cane culture. Beyond these fields are long lines of barracks

or "quarters," painted white, and flanking the clustered stone and brick buildings of the estate. A
flavor of sugar, and a vague suggestion of something
more potent hangs over the whole place. There is a
moist freshness and greenness in these big cane fields
that one seeks for in vain elsewhere in the tropics.
An anonymous author, in a recently published paper
in Blackwood's Magazine,—one from whom we may
quote again, as his statement of the condition of affairs
in Jamaica seems the fairest and truest of any recently
published, says:

"Now, as heretofore, the most important agricultu-
ral produce of Jamaica, as far as the outside world is
concerned, is the cane, and sugar and rum are the
largest exports. In most districts in the island the
eye is at once attracted by the stretches of emerald
green cane-pieces, and, in the centre of each estate, by
the tall chimney of the *engenio* where the crop is
crushed, the juice is operated on, and the refuse of one
manufacture forms the basis of another. The halcyon
days of the sugar industry are past, and the profits of
old times may never be reaped again; but even under
present conditions, the sugar-cane crop pays well, and
capital coupled with energy and industry have, as we
said above, no reason to be dissatisfied with the result
of their efforts."

The last fiscal report shows that there are a total of
32,515 acres of cane, under cultivation in Jamaica, a
falling off of nearly one quarter during a decade.
This reduced cultivation is said to be caused by the low
price of muscovado sugar in the European markets.

The Albion estate, considered for many years one of the best and richest in the island, contains 4,976 acres, only a few hundred of which are now under cultivation. All the latest improvements in sugar machinery are in use there, vacuum pans and centrifugal process; yet the other day it went to the hammer.

Leaving Albion with its living green behind us, beyond the Yallahs River, we drive into the picturesque straggling little town of Yallahs, around a turn in the road and up to Mother Noel's house. A good sized two-story frame house it is, with a steep stairway at one side, by which you ascend to the entrance on the second floor. Mother Noel herself meets us at the door and conducts us into a neat parlor or hall, which occupies the centre of the house, and from which four rooms open. There is a passage at the front, going clear across, and another to match it at the rear; from either of these the main room is entered. This apartment contains some rather good old mahogany furniture, solid, of course, as all furniture is in this country, where mahogany may be had almost for the cutting. Mother Noel, now in the neighborhood of eighty years, has passed much of her life here, and is widely known to travellers as one who keeps a contented mind, a good reputation and an excellent table. There are few lodgings, in Jamaica or elsewhere, where one can procure a good cup of coffee, but this is one of the few.

Beyond Yallahs, after passing the salt ponds, the country is full of streams which one must ford, the water often coming up to the wagon hubs. There are stories told of different people who have been caught by the sudden rising of these waters. A sailor, and later still, a postman, lost their lives in trying to "do" the fords when the rivers were "down." Finally we reach Morant Bay, a small place where there is little accommodation for the traveller, but much to interest one, both in the natural scenery and sea view and in the large shipments of fruit made from here. Besides this, there is much that is interesting to the student of history in Morant Bay.

We could not find lodgings, the one room devoted to that purpose in the town being already pre-empted; but we were afterwards rejoiced that it was so and inclined to attribute it to good fortune, as we were kindly entertained by a gentleman of the neighborhood, from whose house on the hill all that is most attractive in the neighborhood is included in one comprehensive view.

It was here at Morant Bay that the first scene in the tragedy of October, 1865, was enacted, as already related in the second chapter of this book. The vestry of St. Thomas ye East, met at the court-house at Morant Bay for the transaction of parochial business. At three o'clock on the eleventh day of the month, several hundred people, crying, "Color for color," closed in about the building and began to stone the volunteers, who were drawn up to guard the members of the vestry. The story has been well told in the following concise language.

"The Riot Act was read and the Volunteers fired, but they were soon overpowered. A hand-to-hand struggle ensued, during which Captain Hitchins, faint from the loss of blood, rested on the knee of a Volunteer the rifle he had taken from a murdered comrade and fired his two remaining rounds of ammunition. He was then surrounded and hacked to death. All the officers and many of the members of the Volunteer Corps nobly died at their post, gallantly doing their duty. The Custos of the Parish, the Curate of Bath, the Inspector of Police, and a number of Magistrates and other personages were also murdered."

One who journeyed through this same region only a few years ago, spoke of the clouded looks and morose expression of the negroes, who seemed to remember the terrible chastisement which followed this outbreak, and to be waiting their chance for revenge; but to-day there seems to be nothing left of this feeling. When a man was starving, or next door to it, waiting his chance for a job which would pay him a shilling a day, he could afford to remember a smart; but after he has found it better to work than to brood, he begins to forget past grievances. Fruit growing and other blessings that have come in its train have gone far to make a contented peasantry, and to draw a veil over the horrible events of less than a generation ago. Although the buildings were nearly all burnt at the time we have written of, there are now a hospital, almshouse, court-house and constabulary station, an iron market building, a post-office and telegraph station, an Episcopal church and one belonging to the Wesleyan congregation. This for a town of little

more than one thousand people is not a poor showing.

Bidding farewell to our good host at Morant Bay,
we approach Port Morant, where Captain · Baker's
estate of Bowden is, and from which a quantity of fine
bananas are shipped by the various companies engaged
in the fruit business.   It is interesting to see the
great vessels of the Atlas, the Boston Co., and other

steamship lines come into this quiet, enclosed harbor
and transform its repose into activity.

But we are going on a little too fast.  We pass on
the way woods, where it is said that the hero of old
time melodramas, " Three-fingered Jack," performed
his feats of derring do.  Of course the searching eye of

modern investigation has reduced Jack to the vulgar proportions of a brutal negro foot pad, whom we are glad not to meet nowadays.

The Morant River has quite a delta, and at its broadest branch we stop to watch the women who are washing and gossiping just above us.

There is a riding road along the Morant River, to a place in the interior called Island Head, in the coffee region. From Island Head, a bridle path will take one, by the course of an old road built by Governor Trelawney, in the last century, but now gone to ruin, over the mountains and into the Maroon settlement of Nanny Town, named after their once notorious chief. We will have more to say about this place in another chapter.

On the way to Port Morant we must not miss the view from the turn in the road above " White Horses," a cliff which makes a prominent coast mark to mariners, and whence a grand ocean view with foreground of picturesque rock and enchanting verdure makes the traveller long to pitch his tabernacle there.

The nearer approach to the shipping place of Bowden is first through the village of Port Morant, a little cluster of houses and cabins around a cross road where some great trees throw their shade ; beyond whose trunks are vistas of white road, thatched roofs, palm tops and streams. Then the way skirts an unsavory morass, into which the tide flows among the mangrove stalks. After that a sharp turn, and it follows the curve of a hill base, passes a little settlement and ends at the storehouses and wharf.

From here, as from the hill top, the view is wonderfully fine and the air all that could be desired. In spite of the swamp, the people seem to be blessed with abundant health. Bowden is one of a number of estates which, no longer valuable for sugar and rum producing, are adding their quota to the new wealth of fruit and cocoanuts.

Many people are bringing down the bananas from their little 'fields, back in the hills. They are put into the storehouse in open slat crates or bins, where the air has full access to them. All the fruit is picked green and shipped in that condition, and when placed on board the steamers is so arranged that each separate bunch is well ventilated. This is done by building open stalls of slats between decks, or hanging the bunches in tiers. Great canvas funnels are put down through the hatches whenever the weather is sufficiently fine to allow the latter to remain open, and these, with their broad mouths stretched open to catch the breeze, carry it into the hold.

# BATH AND MANCHIONEAL.

THE country is richer and more tropical as we leave the coast and drive up towards Bath. At Port Morant, or near that point, the road makes a sharp turn to the north, and the impression of lavish expenditure of energy on the part of nature is heightened at every step. The road in the neighborhood of Port Morant used to be famous for nothing so much as its mud. People stuck there frequently; especially was that the case with the heavy wagons loaded with rum and sugar, which now travel easily on one of the best roads the island affords. There are twenty miles of this highway where you will not find a break, or any unevenness or mud. It is as perfect as possible, and £500 per annum keeps the whole twenty miles in repair. When you consider that this road is solidly built, with stone culverts, bridges, rock terracing and hill work all the way, the result surprises you.

The negroes who work upon the road are small contractors. Work on concrete, iron work or what not is done at so much per yard, and the workers earn from one to two shillings per diem, usually knocking off on Saturday, reserving that for market or field work.

As we advance we lose the bits of marine view that added so much to the beauty of the Windward road.

These views are replaced by no less enchanting glens
and ravines, into which the rich, deep, pervading, all-
enveloping fulness of sylvan life floods like a tide,
overshadowing the road and rolling in billows of
verdure up the hillsides. Along such a way, past
a curious hillside with vertical strata; marking the
end of a moraine and making a note of interroga-
tion for those who would limit glacial action to
the continents; stopping at springs where the abundant copper ore told the mountain's secret; crossing
bridges and driving through fords, we reached Bath.
Bath,—once fashionable, but now only occasionally
visited; Bath,—where the lotus must have been imported with the other rarities, upas, cork, gamboge
and such like trees, in the old garden, so that the traveller sits him down in Mistress Duffy's parlor and is
straightway content to let the world outside go as it
will without him.

May we speak of G., who tarried to show us the gardens and sanitarium and to do the honors for the hills
and watercourses? or of Mrs. S., whose collected pamphlets were such an aid to us, and whose garden with
its flowers and fruit and chances for entomological
research was such a delight? or of S., the director whom we met at the bath, whose attentions
were so helpful to us? These were only a few of
many who added to the enjoyment of our sojourn.

The baths, which are about a mile and a half from
the town, which owes its existence to this proximity, are
at the end of a winding road bordered with vines and
moss and fern covered rock, flowering shrubs, trees

heavy with fruit and an atmosphere charged with moisture and very fragrant, like that of some vast greenhouse whose temperature and humidity had been regulated to force vegetable growth' to the utmost limit of its possibilities.

At the bath we find a building in charge of a one armed man, who introduces visitors to the stone basins built to receive the hot and cold water that flows from

the hillside within a few feet of each other. There are in Jamaica, as noted elsewhere, several baths of a medicinal nature. The one we are visiting has been chiefly valued for its unquestionable influence on rheumatic and cutaneous disorders.

Much has been written of these springs; nearly all the historians of the island have had something to say of the "baths," as they are commonly called, though only those at Bath, in St. Thomas, and Milk

River, in Clarendon are so fitted up with houses, etc., as to deserve that name. The only other spring that seems to call for special consideration is that known formerly as the "Jamaica Spa," in the St. Andrews' mountains, near Newcastle, but long ago abandoned and left to go to decay.

The Bath of St. Thomas is derived from a sulphurous sodic calcic thermal spring; that at the Jamaica Spa from an acidulous ferro aluminous spring; and that at Milk River a saline calcic thermal.

Straying further up the stream that brawls by the station we were suddenly caught in a most terrific downpour of rain. The broad leaves of the wild plantain or the shadowing limb of a bread fruit tree made but poor shelter. But soon the sun came out to cheer us and the great greenhouse of nature steamed again, till it seemed as though the expanding of trunk and leaf occurred visibly.

Besides at Mrs. Duffy's well kept house one can find lodging, though no board, at the bath building.

Bath was the scene of some of the great atrocities of the insurrection of Governor Eyre's time. Some of the people of the village still have tales to tell of the pillage of valuables, of sudden flight to the woods, of hardship and of suffering. The town is a small but very pretty place. It has its church and chapels, court-house and constabulary station, and a small population who are most contented and industrious blacks.

From Bath one can go by a bridle road up to the weird and wonderful "Cuna cuna" pass in the Blue

Mountains, a ride of rare beauty and interest, and from
thence descend through the region of the Rio Grande
to Port Antonio, past the Maroon settlement at Moore
Town ; or he can, by following a road that skirts the
Plantain Garden River, reach Island Head and
from there, by the way already described, to Nanny
Town.

All of this ridge, and the country north and

east of it, is of great interest to one who enjoys a little hardship with his travel, for it is an unsettled and
untravelled country for the most part.

We must not neglect to speak of the Scotch gentleman, engaged in coffee planting, whose optimism
threw a rosy light over the sable toilers of the land.
He says that the labor problem here is only a question of fair wages.  In his opinion the black man is

not lazy, but underpaid. The superintendent of that piece of road we have admired adds that he can always get good laborers by paying living prices. " Do not fine your men. When they don't suit, discharge them ; when they do, pay them."

One of the characters of Bath is a soldierly old fellow who modestly hides a Victoria cross, except on pension days. He stood with the little party that Havelock relieved at Lucknow, in India, and now he bosses a gang of negroes on the highway, in Jamaica.

From Bath to Manchioneal the way is more level, passing bottom lands that are frequently overflowed, and meadows that are like those of the mother country. At Manchioneal we tarry at a dingy lodging, " not too bad " and certainly not too good, within sight of the sea. To the right, are the clean, bright looking buildings of the constabulary station and the church. To the left, a high bluff hides the road along which we are to journey.

# IN PORTLAND—PORT ANTONIO.

Of Damascus the ancient prophet wrote, " It shall be a ruinous heap." Did any one prophesy concerning the great estates that lie along the road from Manchioneal to Port Antonio? In the quaint, terse language of the courts, they are " in ruinate."

In the parish of Portland in the county of Surrey, between the John Crow Mountains and the sea, lie many abandoned estates; buildings, walls, chimneys, aqueducts, all going to pieces, and the oncoming tide of foliage, like a green wave, engulfing them. There is much of beauty and interest in this eastern end of Portland. The road winds with the turning of the coast line, and constantly affords surprises and scenes of rich beauty. Deep bays and inlets, beaches where the transparent water breaks in a long surf, headlands crowned with foliage, all afford satisfaction to the eye. Innis' Bay, is a deep indentation in the coast, upon which we come suddenly, having crossed previously a bit of uncultivated land shut in by bushes and trees. We look down from above on the expanse of emerald water and the worn rocks, white with its foam. Tropical trees frame the picture. Here are the Fairy Hill Bay, with its extensive outlook each way over the ocean; Priestman's River, deep at the mouth as it debouches into its little harbor; and the exquisite " Blue Water," whose turquoise shades into amethyst in the shadows, over which the bending trunks and swaying tops of a hundred cocoanut trees cast their reflections. All through this region are scattered scenes of rare beauty. At intervals, we arrive at pens where cattle are grazed, and estates that are converted into pasturage for horse kind. Little hamlets here and there, scattered along the road, show a healthy-looking but meagre population. The country, like many other neighborhoods in Jamaica, strikes one as being under populated—which indeed is the case.

But in spite of its beauty, its natural fertility, its advantageous situation, its grazing pens and villages, Eastern Portland gives the impression of desolation. Mile after mile of unused, unredeemed acres, once flourishing with cane, but now given over to wild growths, sadden even the most optimistic observer. Here has been dreadful loss. The cause of this desertion of estates has been already noticed in the earlier chapters of this book.

But a surprise greater than any of those that have greeted us upon the road, is in store. There comes a point where the decay is arrested, and a new life appears to animate the scene. The population is larger and thriftier, the waste acres are taken up and planted with fruit. Everywhere one sees increasing evidence of greater prosperity. Why is it? We have seen how king cane was dethroned : now we are present at the coronation of king banana. " Lo! the old order changes, giving place to new.".

We drive into Port Antonio and up to Mrs. Brown's
lodgings ; a large, well and cleanly kept white house
on a hill, from which we may overlook part of the
town and see one of the harbors. Port Antonio is
blessed with two harbors, divided by the rocks of
Navy Island. The western harbor is the greater :
here large steamers can lie close to the wharves.

Properly there are two towns, upper and lower
Titchfield, the division being made by the hill.
Lower Titchfield, or Port Antonio, lies along the
beach on the lower land, and contains the governmental and mercantile buildings. The hand-book gives
this useful piece of information and advice to mariners approaching Port Antonio. "The fort and
barracks are conspicuous objects from the offing.
Navigators strange to the locality, sometimes find it
difficult to distinguish the entrance to the harbor, and
if a vessel should approach the shore to the eastward
of it, the remains of some old sugar works in ruins
might be taken for the old fort at Titchfield and
prove misleading, but by running along the land the
place, when once opened, cannot be mistaken. The
new light house on Folly Point at the entrance of
the harbor is a great aid to navigation."

In the year '68, which has already been spoken of as
one of remarkable promise for Jamaica, the initial effort
was made in fruit shipment, which has resulted so ben-
eficially, not only for Port Antonio, but the whole
island of Jamaica. The author of " Picturesque Jamaica," refers to the pioneer banana shipper in the following terms. "Above fifteen years ago, a Yankee

skipper knocking about with his schooner, had occasion to call at some of the ports on the easterly part of the island. His keen eye looked with interest on the bananas that were so plentifully offered him, and knowing the taste the Americans were fast acquiring for this delicious fruit, but which was rarely found in the American markets, set himself to the task of devising means to convey the fruit in a sound condition to those markets. The success which has followed, is shown by the fact that the shipment of bananas to America has become one of the leading industries of this island." The Yankee skipper referred to—now more commonly known as the Banana King—is Capt. L. D. Baker, whose interests, merged in those of the Boston Fruit Company, are to-day among the largest in the land for which his enterprise has done so much.

One of the prominent objects seen from the road to the eastward is the Episcopal church, a building of some beauty and uncertain age—or rather, having a record of moderate age duly preserved, has gained the popular reputation of being ancient. Like many of the " very old " buildings now standing in Jamaica, its pretensions dwindle when examined carefully. Quite a young man, whom we questioned, assured us that it was standing there *when he was a boy.* Port Antonio has rather a fine court-house and jail. On the hill are the Titchfield Trust (school) almshouse, and residences.

There is in the town quite an American population; greater in proportion to the size of the place than that of any town in the island. The business at Baker's as well as in some other places, is conducted by Americans, and there is quite a New England air about the place. To appreciate that, however, the visitor must first have seen some other parts of the island, and have become accustomed to the different tone which pervades the south side. Port Antonio's population is about two thousand souls.

The hurricane of 1880 did a great deal to damage the fruit interests of Portland, but it soon recovered and redoubled its enterprise.

There is, thrice yearly, a fair for the sale of stock in Port Antonio, and semi-weekly markets are held in the substantial building provided for that purpose. The Wesleyans and Baptists have each their place of worship in the town, and these are well attended. Many of the houses are substantial and picturesque and the climate is delightful. Down about the wharves, where the fruit is brought in mule carts, drays and on the heads of the native black people, there is a considerable activity. Mr. Moodie, the representative of the Jamaica Fruit Company, whose office is here, assured us that there are few of those who may be seen bringing the fruit into market in this way, who have not a little store of money or a bit of property,—provision made for a rainy day.

Says a writer already quoted, writing of this same port : " If the people of Jamaica choose to exert themselves to supply this demand, an era of prosperity, unknown in its history, awaits the island in its near future, and Jamaica might become the tropical garden of America." Over a limited area this prophecy is being fulfilled.

Over the mountains winding down the road leading into the country from Port Antonio, drops us
into the rich valley of the Rio Grande. It is a road
"not too good," as the negroes would say, yet better
than most mountain roads in the United States. It
is encumbered with some rocks and inequalities, and
beautified by many windings. The country through
which it passes is rich and fertile, well cultivated, and
abundantly blessed with picturesque views and color
bits. The Rio Grande, rising near Bath, twenty-five
miles from the sea, flows through the heart of the
banana country. It receives tributaries from the north
side of the Blue Mountain peak. It is the second
river in size in Jamaica, and one of the swiftest of those
erratic streams that flow pleasantly within narrow limits one day, and the next sweep down, full and terrific torrents, angry and swollen by the rain-fed streams
from the surrounding mountains. At one point on this
river, a place where men with great bunches of ripe
fruit, donkeys laden with well filled panniers, and even
children carrying "pick'nies" smaller than themselves,
cross and recross with safety, the erosion of the banks
shows that not infrequently the wide stony bed must
be covered with a flood several fathoms deep, where
ships might ride if they could stand against so powerful a current.

All about Jamaica the waters of the rivers rise, or,
as the native says "come down," (i. e. from the mountains) very suddenly, and often travellers have been
imprisoned for days between two torrents, on a strip
of country when there can be found neither town

nor lodging house. At such times he trusts to the ever ready hospitality of the Jamaican Creole.

Along the Rio Grande we find several of those conditions which are commonly counted among the advantages of Jamaica. There is sweet, pure air, and plenty of it ; a superabundance of clear sky ; a sufficient rainfall ; good soil, and an industrious and contented peasantry.

Crossing the river by a ford from which a most bewitching view of mountains, wooded point and level mirroring pools may be enjoyed, we soon enter the property known as " Golden Vale," once a great sugar estate, but now converted to banana cultivation, under the enterprise of the Boston Fruit Company. This company has bought and leased a great many estates on the north and east end of the island, and its promoter and manager, Captain L. D. Baker, is looked upon as one of the greatest benefactors the island has known, through his wisdom and energy in introducing, fostering and conducting the immense banana trade of Port Antonio which, within a few years has revolutionized the trade, prospects, and even the habits of the people of this section.

Golden Vale has an output of from twenty-five to thirty thousand bunches annually. It is only one of nearly a dozen estates belonging to Captain Baker's company, yet in the total amount of fruit shipped by the steamers of this line the proportion from its own cultivation is only about one-fifth. That is to say, the people, the peasants with small holdings on the moun-

tain sides and along the roads, are doing what they
have never before had a chance to do in the "Land of
Streams," and are independently working out their
own salvation.

It cost, we are told, about two pounds per acre to
clear new land, and there is little in the market at any
price, so that most of the holdings are leased. Since
the introduction of the banana industry, properties
which before could not be disposed of at any price,
are now scarce and in great demand.

It may not be out of place just here to describe the
growth and appearance of a banana plant, for the benefit of those who have only seen the delightful fruit
hanging in fruit or grocery stores. In general habit
and growth the banana and plantain can scarcely be
distinguished from each other. Both have a soft
stalk, from four to eight feet in height, spreading out
at the top in a cluster of great broad leaves, often
fifteen inches in breadth and four or five feet in length,
bright green, and translucent. When first unfolded,
the leaves suddenly unroll, sometimes with a hissing
report, and appear with unbroken edges and unsullied
surface, across which the shadows of other leaves fall,
and with which the wind toys. But in a very little
while this play of the wind has split the great surface
into a hundred transverse streamers, all attached to
the central vein. Each plant bears one bunch of fruit
which hangs with the "hands," or separate sections,
curving upwards, the reverse of the fruit store method ;
and from the end of the bunch, on a short green stem,
is the blossom, a great heart-shaped, maroon colored

plummet, about the size of a man's fist, or larger. The plant bears only one bunch of fruit.

In planting bananas they are placed fifteen feet apart, and when a plant has finished its mission it is replaced by a sucker.

At Golden Vale the whole landscape is one of rich, perfect cultivation : the fields of cane grown as fodder for the numerous cattle used upon the planta-

tion; the herds of oxen and droves of mules. Beyond the cane field are acres, hundreds of them, emerald with the ranks of bananas. Not far from the boundary of the old estate are the great stone buildings formerly used in the crushing of cane, the manufacturing of sugar and rum, storage and the preparation of indigo. These are now converted into shops, depots and school-houses. Most of the children in the neighborhood of Golden Vale attend this free school, which is kept up by the bounty of the owners of the plantation, and it is a very pleasant sound to hear the hum of recitation, or the melody of song, from the grim old walls, where sounds of a very different nature once were heard.

Upon the ruins of very extensive buildings near the top of a little hill stand the houses where the Busher (overseer) lives and directs. Near by, across a small valley, is a settlement or barracks, where the coolie laborers, of whom there are a hundred or more on the plantation, live in their unobtrusive way.

Some national or religious holiday was being observed when we were there, and several of the men were engaged in decorating a little building like a doll's toy house with bits of tinsel and colored paper. This was to be used in some ceremonial. They were very polite in showing us this wonder, but we were warned not to touch it or to show anything but respect.

We inquired respecting the prices paid for labor on a banana plantation, and found that a reaper might earn with industry seven shillings a day, while the trimmers and other laborers range from one shilling and sixpence to two and sixpence a day, the lowest price being paid to the women.

Standing on the piazza of " Busher " Davis's house, and admiring the perfect view, we were soon informed that horses were ready for us. We mounted, and following our leader, crossed the little stream that crosses the valley just beyond the buildings, and struck into a trail which led us into the hills. Higher and higher we mounted, finding at every turn more extensive views and a more exhilarating air. New fields, fresh cultivations, unimagined richness disclosed themselves to our eyes, as we wound in single file around the shoulders of the hills. Palm trees interspersed with the bananas. Bamboos hedged the emerald fields with their plumes. On the distant hill that lay between us and the slope on which is Port Antonio, there was pointed out to us a cut in the trees where the surveyor's line had been run in planning the projected cable line by which, in future, bananas are to be transported to the shipping place.

Finally we got on a sort of hog back ridge, where the soil was a deep red clay, and the out-cropping white limestone contrasted brilliantly with the earth and the plentiful foliage. At places where the ridge narrowed we could look down into a valley on either side, in one of which flowed the Rio Grande and in the other the Back water. We stopped to steal (?) mangoes from the never failing trees, to inspect the hillside garden, placed at an angle of forty-five degrees, where some peasant has raised coffee, pimento, yams, potatoes, Seville oranges, plantains, guava and a dozen other things. We must accept a " rose " (really Indian Jessamine) from a pretty peasant girl, and look at the boars' teeth which a lad in one cabin showed us, pointing with pride to his hound, by whose aid the wild boars were killed—" Over dere in de John Crow hills w'ere me daddy hunt 'em fer two day, sah."

GOLDEN VALE BANANA PLANTATION.

Still upward the path goes. It is a precipitous trail. Inquiring how it came to be made, we learn the interesting fact that it was trodden first by mules, bearing copper ore from a mine several miles beyond the summit. The Golden Vale mine was in operation thirty years ago, and some good copper was got out ; but the distance from the market, and, it is said, a quarrel among the members of the company resulted in its abandonment.

It will not be wise to close this chapter without more definite reference to the work and influence of the Boston Fruit Company, whose efforts at building up the fruit trade, in conjunction with the Atlas Company and the Wessels Line, have made such an important difference in the condition of trade upon the north side of Jamaica. We make this mention here rather than in the chapter devoted to travel and communication, because the interest in the company centres here in Port Antonio.

The properties of this company are in many cases tenanted by the negro peasantry, whose industry is stimulated, as we have already pointed out, by the fact that they are actually laying up money. The establishment of many such schools as the one at Golden Vale, where the children of these peasants may be taught free, is one of the methods by which the company is seeking to improve the condition of its tenants. The tenantry aggregate about twelve hundred souls. This, of course, includes all in different parts of Jamaica, not alone those near Port Antonio.

Following is a list of the properties of the Boston

Fruit Company: Bowden, which we have already visited at Port Morant, and which is not only important as a shipping port, but is also a fit place for a sanitarium; its acreage is about four hundred. Philipsfield, Williamsfield, Unity Vale and Elysium have each 1100 acres. Seamans Valley has 1000 acres. Hermitage, whose output of nuts we have especially mentioned, contains 560 acres. Wyant, Upper Layton and Red Hassel are the smallest of the principal properties, comprising only about two hundred acres each. Size is only comparative after all. The man who has 200 acres of good land in New England is a large land owner. Wentworth, Lookout, Fellowship, Prospect, Hermitage, Windsor, Paradise, Wheelerfield and Plantain Garden estate are properties that range from 450 to 900 acres each. Fairy Hill and Bound Brook each contain 1800, and Golden Vale rises to 3500 acres, with all its dependencies. It is a noble aggregate.

From the Hon. W. Bancroft Espent, the company has leased Spring Garden estate and four others. Byram is a beautiful piece of fruit land. There is here a railway for tram cars, the only railway privilege on the island except that of the Jamaica Railway Company. This extends three miles.

The shipping places of the Boston Fruit Company are at St. Ann's Bay, Ocho Rios, Oracabessa, Port Maria, Annotto Bay, Buff Bay, Orange Bay, Hope Bay, St. Margaret's Bay, Port Antonio, Blue Hole, Manchioneal, Port Morant, Morant Bay and Yallahs Bay. At each of these places the company own properties.

The Boston Fruit Company run six steamers between Boston and Jamaican ports. They are dispatched semi-weekly. Their principal use is the carrying of fruit, but they have also each passenger accommodation for twelve people, the rate of fare being $50 for round trip, or $30 either way. As stated, these vessels make stops at all island ports.

# THE DISTRICT OF ST. GEORGE.

WE learned from Mr. Moodie, the representative of the Jamaica Fruit Company and one of the earliest handlers of bananas on the north side, that the fruit land in the neighborhood of Port Antonio is being eagerly watched for and bought up. Buying large quantities of fruit for shipment to "the States," he says that the supply is increasing rapidly with the growing demand. "And," added he, "don't think these people are poor because they are poorly clothed. Go to their churches, or, better still, attend one of their weddings. You will find the men in broadcloth, and the women in silks. Few of them are without little bank accounts. They will improve with their fortunes. The man who works with a machete, in rags and tatters, rides a good horse and carries his silk umbrella on a holiday. The same class of people cannot do that outside of the fruit districts, and these facts answer the question whether they make more than a bare living here. Good clothes, better houses, savings bank accounts, all mean that there is enough and a little to spare."

Which argument seems to us forcible. But let no reader of this rush to the other extreme of imagining that the negro has attained perfection because he is partially emerging from his former brutish condition. He is still full of faults, given to all manner of unreasonableness, and prone to leave you in the lurch if you are obliged to depend upon him. He has still very hazy notions on many moral and social questions. He will sometimes provoke you into wondering why any sane person should have deprived his mother country of him. Then you notice that he begins to be self-respectful, ambitious for his children, more conformable to law, cleaner in his house and living. To the question : Whence is this improvement? the answer is ready. It is threefold : A better government, freer education, and living prices for work.

The Maroons, who long ago wrested a sort of liberty from the government, and who have lived for a great many years unmolested in the hills, enjoying certain privileges and immunities, come down also to trade. It is too early yet to predicate any sudden or immediate change in their condition, but it is safe to say that with growing intelligence and prosperity, these people will become gradually merged in the common population. Nothing is so certain to break away barriers than an advantage on one side of the fence or another.

Mooretown is now the nearest Maroon town. To reach it one must follow the same road that goes to Golden Vale, which is about half way out to the village. In the village live many of the principal Maroons. Further up the road the wild and beautiful Cuna Cuna pass is reached. Only on horseback can one advance as far as the pass. Having crossed it and enjoyed its coolness (and perhaps a sudden shower as

well), the traveller descends by the bridle road that
we have already used to Bath.

The western coast of Cornwall, beyond Port
Antonio, is as fruitful as the eastern is desolate. At
some points the road is lost in a forest of cocoanut
trees.

The largest cocoanut grove on the island is that of
Wentworth, the weekly output of which is over 20,000
nuts. The Hermitage, at which we halted a few
miles west of Port Antonio, is about one fifth smaller,
being the second grove in size and productiveness in
Jamaica. This is also one of those properties of the
Boston Fruit Company whose immediate details are
usually managed by intelligent, active young men of
New England blood. In this case it was one of Captain Baker's sons who did the honors of the place,
showing us the (to us) wonders of cocoanut growing
and the natural beauties of the neighborhood. We
drove out to a very pretty and airy little iron bridge,
whose light arc crosses the Swift River in a single
span. Then back to the ford, where we parted reluc-
tantly from our young guide.

Perhaps (if the old resident will be patient with us
for a few moments) it may not be uninteresting to
those who have only seen the cocoanut in market to
know that if they were to see the fruit growing they
would not recognize it, probably. Indeed, it would
take sharp eyes to look through an inch or two of
enveloping shell and find the familiar nut hidden
beneath. Picking cocoanuts should never, so say
the experts, be allowed. A tree here takes seven or

eight years to mature; but when it is of age it produces clusters of nuts· that are all clinging with their
little separate stems to a common stalk. A tree may
have a great many clusters growing at one time, from
the very small ones to those that are ready to fall.
For some reason not yet known to science, the bearing
capacity of a tree is injured by the picking of its fruit;
so that only the fallen are gathered. Then the work
of husking must be done before the nuts can be
shipped. This is done with a machete; the heavy,
fibrous, yellow outer husk with its pointed ends and
three dull angles is cut away, and the cocoanut, greatly
reduced in size, is marketable. It contains at this
stage a quantity of the pleasant fluid known as cocoanut water, and the firm meat is much softer and more
delicate than is usually the case with the older nuts
that have got to the Northern markets. To get
"jelly" nuts, whose contents can be eaten with a
spoon, they must be picked from the tree before they
are quite ready to drop.

While we are driving through this pleasant country,
it may not be out of order to speak of another matter
quite germane to the purpose of this book, and consider what the chance is for enterprising young men.
The anonymous author in *Blackwood's*, already quoted,
and generally sensible in what he has said on this subject, makes a statement and a suggestion which does
not, it seems to us, accord with wisdom.

"There is every ground for reasonable belief that
Jamaica now offers a remarkably favorable opening to
the numerous class of young men, in England, who

are unable to pass the examinations which are necessary for entrance to the army and civil service, not to
speak of the more learned professions.　Young men of
this class have lately been shipped off to Australia,
Africa, and America to seek their fortune, sheep-farm-
ing, gold-digging, and ranching ; and we leave it to
those who are interested in them to say whether they
have, as a rule, gained fortunes, or made more than a
livelihood, meagre out of all proportion to the rough
life which they have led, and the toil which they have
encountered."

That is what Jamaica does not want.　England has
often shipped black sheep, and lame sheep out here,
but they have not been the successes to which their
countrymen usually point with pride.　The only kind
of a man who can expect to *live* in Jamaica is the man
of clean habits.　There is no climate that will kill off
rakes and rounders faster than a tropical one will.
The only one who can *succeed* here is he who exercises
the same faculties that lead to success elsewhere.
Only clear-headedness, pluck and habits of work will
go further here than in most countries, as the land is
not yet overstocked.　There is a call for capital; there
is a field for energy; there are opportunities in the new
state of things that will make the intelligent investment of either one profitable.

Richard puts a period to these reflections by pulling
up.

" Dis here is Lowlayton, gentlemen, what you
wanted to fine."

Lowayton, we have been told, is the place where

the traces of an extinct volcanic crater are to be found. The information was correctly and judiciously worded: traces of an extinct crater, certainly, inasmuch as that which has evidently been thrown up by a volcano is good evidence that the volcano has at some time existed. We have also every reason to be satisfied that this volcano has been very extinct for some time. However, as this is the "only volcanic formation" noted on the island, we must e'en make the most of it.

Near the Spanish River we halt and so-journ as guests at a house on the hill-side, the courtesy and hos-pitality of whose in-mates would launch us into description and acknowledgment, did they not more strongly compel us to silence.

In this neighborhood is much to excite inter-est. The sugar producing will here soon give place to fruit growing. The hills produce logwood, and coffee is raised in the vicinity, we understand. Nuts too are quite plentiful. The hills are full of delightful nooks, and woods, and watercourses, for the lover of nature. On one occasion we rode along a bridle path to its end, and then, following a sturdy negro, who went

slashing with his machete through the brush, we came suddenly upon one of the most perfect little streams that ever an artist imagined. Its beginning (for us), was a cascade of some sixty feet in height, that shot like an arrow from the fern-covered rock crest to a circular, clear pool, whose further margin had burrowed away and hidden under the base of the rocks, and around which all manner of trees and wild plants grew. The cotton trees reached out their great buttresses towards it, balancing the heavy spread of foliage over-head. The wild tamarind and the locust, or, if not these, some other equally native and beautiful trees, stretched out their arms and dropped long branches of vine, like plummets, down to the water.

From here the brook goes, like any brook in any other land, doing all the usual erratic and charming things that any other brook does, till it goes to join some larger stream. There are moments of dignity: pools placid and calm, but just over the edge of them the frolic and rush begin again. If among our readers there is a poet who cares for a suggestion,—?

Over the Spanish River a new bridge was in process of construction. The honest work being bestowed upon it, and its solid character, are in keeping with most of the work of this class done in the island.

# FROM BUFF BAY TO THE ROARING
# RIVER.

CROSSING several little streams and an occasional
hut or hamlet by the wayside; enjoying the glimpses
of the mountains in the distance, or the verdure of the
hills near at hand ; pausing at the fords to water the
horses, and to exchange badinage of an elemental
character with the half aquatic women, we at length
arrive at Annotto Bay, where it is proposed to satisfy
a growing inward yearning after the flesh pots.

Having ordered luncheon at the principal lodging-
house in the town, a little dingy looking place nearly
opposite the market and bank, we strolled back to
inspect the bridge being built over the Annotto River.
A constant stream of people were passing over the little foot bridge below the ford. Among these were
many coolies, more than one usually sees even upon a
market day. White turbans, close fitting skull caps,
loose upper garments and bare legs, faces like those of
Europeans, intellectual and dignified, bearded and
venerable, thin, wiry bodies and erect carriage ; so the
crowd of coolie men go by, returning our salutations
with grave courtesy. A military salute, with the fore-
finger touching the forehead and the palm turning
outward with the completed gesture, and the one

word "Salaam," seriously spoken. The women, of whom we have seen many elsewhere, are also present in force, and so are the negroes of all grades.

Up and down the river, from the framework of the new bridge, we saw the bending plumes of the bamboos and the green of cultivated fields backed by the hills. It would not have been difficult to imagine ourselves in a suburb of Benares. The sunshine became broiling hot. We stopped to watch some men trading and selling horses, and then started on a tour of inspection of the shops: it being Saturday, they were thronged with people buying what the crowded market could not furnish them. A dry goods store, so full that one could hardly turn around in it, was an epitome of Macy's or Wanamaker's. In a drug store there was a man dealing out spirits in the rear and another in front selling cheese.

The market place was filled with the people, buying, selling, chaffering and making a medley of sound that did not present one intelligible word to the stranger. Squatting on the ground, seated on benches, standing in corners or along the fences, what a motley multitude they were. The wares exposed for sale were tempting (until we had lunched). There were bread-fruit and nuts, yams and potatoes, —sweet potatoes, of course,—plantains, ginger cakes, cassava wafers, crabs, eggs and poultry, turtle (think of them baked and served in the shell), confections, made of cocoanut and brown sugar, and all the various things that can be grown or prepared or cooked to tempt the Creole appetite.

But we were allowed to satisfy ours with more familiar viands. Ham and eggs followed a soup over which prudence prompted to say grace without inquisitiveness ; boiled yam was vis-a-vis to fried plantain ; and the maiden who served the table was brown and inclined to be coquettish.

We had several places in the town to visit and so did not linger.

The town of Annotto Bay is the second in size and importance in St. Mary. It lies near the mouth of the Wag Water : the *Agua Alta,*—the Loud River. A shipping port, it has, in the past, done considerable business, but has latterly fallen somewhat from its mercantile estate. One great drawback to the town is the proximity of marshes which breed fevers and malaria at certain seasons of the year. It possesses some good buildings : the court-house and savings bank, hospital, church, and Baptist and Wesleyan chapels. The new junction road, which we have followed from Kingston as far as Castleton Gardens, meets the coast road at this point. Near here are the Port Maria water works. Were it not for the overflow of the several streams into which the Wag is here divided, and the consequent ill health which strangers are apt to endure in the wet season, there is no reason why Annotto Bay should not be one of the chief ports in Jamaica. And the obstacle is by no means an insuperable one.

From Annotto Bay the road makes a wide sweep, leaving the coast, which here lies northwest and running nearly west to a place called the Cross, where is the junction with another highway which leads to the southward, crosses the Flint River, follows the course of the Rio d'Oro and meets the Spanish Town and St. Ann's Bay road at a point midway between Bag walk and Linstead. For a pleasant three days jaunt one could leave Kingston by the Annotto Bay road, enjoy Stony Hill, the Wag Water and Castleton as we have described, and then, instead of returning by the same way, keep on to the north coast, taking the western road from the mouth of the Wag Water and returning · by way of Bog walk and Spanish Town. The entire trip is not more than sixty-five miles, and there are few outings that promise more of beauty and interest than this run around the block. Over a large part of this route the railway branch will run when completed, and the fact that the trains will first reach the coast at Annotto Bay and afterwards connect that point with Port Antonio, means that the history of the little town is still ahead and not behind it,—a rare good thing for a town, as it is for a man.

Turning northward from the Cross, and passing the mail coach on its way from Port Maria to Ewarton, we arrive, near nightfall, at the former town, drive through a long squalid looking street, which hardly seems to be attractive enough for the chief town in the parish, and pull up with a flourish (it is a little way that Richard has) at the door of a dingy and repulsive looking inn opposite the public hospital. We hesitated about alighting, till assured that there was no other lodging-house open to the public in Port Maria.

We had plenty of time to study the hospital, with its corrugated iron roof and latticed sides. It looked clean and cool, and that is a distinction in Port Maria. It was Saturday night. The streets were full of the marketing people ; occasional drunken sailors staggered into the scene and out again, like supernumeraries that had been displaced ; the altercation over a game of something that was going on in the busy lower room of the inn, waxed louder. There was an agent for something or other who had got the best room the hostelry afforded, so we had to take second and third best. Our supper was served in a dirty back room, and it took both hunger and courage to attack it. We soon got out to inspect the town.

"Puerto Sancta Maria," is not living up to its name. It has a good harbor, wharves and stores. It contains a church, kirk and two chapels ; it is blessed with an almshouse and a fine prison. But it, unfortunately, has also that inn.

Crossing a bridge over the river which divides the town, we came suddenly on a group of buildings that decidedly astonished us. There was the church already alluded to, a solid, substantial stone edifice, with a clock tower. Standing near it is a fine large market building, with open sides and corrugated iron roof. Opposite to these is a group comprising the court-house, constabulary depot and offices. These are structures that a much more thriving and interesting town than Port Maria might be proud of. They were built, we were told, at a day when the place had more to recommend it to the visitor. Interviewing a sergeant of the constabulary force as to the behavior of the population (and as we asked we could hear the sounds of a street row in the neighborhood of our inn), we were told that the arrests made were very few.

"Then you don't arrest people for intoxication,— getting drunk and being disorderly ?"

"No, sah, of co'se we neber arres' a ma'an for a ting like dat." He spoke in a grieved tone, as though we had insulted the force.

The loud talking, liberal profanity and occasional expostulation on the part of the proprietor of the inn, did not lull us to sleep that night. We saw the agent for something or other drown his sorrow at having to stay in such a place, after which he attempted, from the window, to put a policeman, down in the moonlight, through the manual of arms. The cracks in the floor were sufficiently wide to allow us to see who was getting the best of the row below and it was long before we trusted ourselves to the joys of not too clean pillows and the companionship of numerous bed-fellows who were not paying ten shillings a night for the accommodation.

We left Port Maria on Sunday morning.

On the road we asked Richard how he had slept, and where.

"Me? I sleep fus' rate, sah. I sleep in de buggy."

We pass the town of Ora Cabessa, with its many acres of nuts and fruit. There are steamers in the little harbor and sailing craft along shore. It is a rising place, and situated in a healthful part of the coast. Ora Cabessa River flows near it and debouches into a pleasant bay. Larger and more important is the stream of the Rio Nuevo, emptying into the Rio Nuevo Bay, where is situated a town of the same name. Here once was an old fort, built by the last of the Spanish governors in his futile attempt to reconquer the island from the English.

And now we come to the White River. It is the dividing line between the parishes of St. Mary and St. Ann. Before crossing it we take a backward look over the district we have left, and consider its chances in the trade revival which seems to be beginning along this coast. There is little cane grown in St. Mary: most of the cultivation is in fruit and ground provisions. The most startling effect in foliage, probably, that ever greeted the eye, is that sea of cocoanut tops interspersed with bananas that grows at Ora Cabessa. As before stated, the chances for Annotto Bay seem almost brilliant, in view of its situation and probable steam communication with the south side of the island. Port Maria, if it holds its own, will have much to do.

St. Ann, into which we are entering, is the spot to which most Jamaicans point with pride. "Have you been to St. Ann's Bay? Ah! then you will see something worth seeing. That is Jamaica's garden."

That depends upon the eyes with which one sees. To perception attuned to English parks, high cultivation and an air of serene thrift, St. Ann is satisfying. It has all the outward visible signs of prosperity in its rolling fields and the green of its perfect verdure, but the ruggedness and the picturesqueness of the mountainous land is gone. We have left the blue mountains behind us. St. Thomas-in-the-Vale affords but a distant and shifting background. The land has become simply moderately hilly, but is abundantly watered by streams of exquisite beauty.

Here it was the Spaniards settled first. Columbus landed here, at Dry Harbor, and not far from the site of his approach the last Spanish governor, Don Sasi, fled in a little boat from the point known as Runaway Bay. There is a bay just a few miles beyond the White River, westward, on which is a town bearing the name of Ocho Rios. The ch in Ocho is pronounced soft by the inhabitants, in violence to its Spanish origin. *Chereras* it used to be—the Bay of the Waterfalls; a name certainly as descriptive as it was poetic. The present appellation, '' Eight Rivers,'' does not do justice to the many streams that rush foaming down the slopes and cool their boiling little bodies in the sea. The harbor is considered a good one. The trade of the place as a shipping port is said to be on the increase, but its morality suffers from the fact that the English men-of-war visit it in order to give shore leave to their crews.

Near here Don Sasi made his last stand. From Ocho Rios a branch joins the great interior road between Ewarton and St. Ann's Bay.

This neighborhood exhibits the most perfect natural irrigation imaginable. The ramifications of the Roaring River, which lies just beyond Ocho Rios, and the roaring of which can be heard for a long distance, are almost numberless, and are due to a peculiar habit the stream has of damming itself at every possible point. The water is full of lime and silica in solution, and these it deposits in walls which invariably check and deflect its own flow, turning it to the right or left where it industriously begins to build fresh dams and seek new channels.

From the work called "Picturesque Jamaica" we quote a description of the source of this singular stream.

" This river rises, or rather appears, about two miles from the sea, issuing from among rocks in a shallow ravine with a large volume of water sufficient to fill a pipe three feet in diameter, and flows rapidly down a narrow channel to the falls or cascade. The only appearance of water above the head is at the Blue Hole, Rio Hoe and Walton, near Moneague, ten or twelve miles away. The flow of water at the head shows clearly that it is not a spring, or the appearance of water percolating through the earth—like the Bog, a mile or so east—but the visible appearance of a large stream already formed and flowing in an unobstructed channel beneath the surface; and it is a singular fact that the volume of water is seldom affected by either floods or drought, never dry, indicating a drainage of a large area. Considerable accession is made to the quantity of water before it reaches the falls, mostly from subterraneous sources."

Where the road crosses the main stream there is a strong little stone bridge, from which one is startled by a view so strange and unusual that it seems to reverse all the laws of flow and growth in woods and waterfalls. There is a pellucid pool, calm, but flowing with a strong current, in which flourishes a grove of mountain cabbage (palm), anchovy pears, banyans and many other plants, ferns and vines, that usually choose dry places for their habitat. Roots form bridges here and there, by which the pool may be crossed at its

Spanish Bridge
Mammu River

lower end, and standing upon one of these, looking up at the cascades, which seem fairly to drop from the branches of the further trees, it is difficult not to give way to the impression that enchantment has been in some way responsible for this strange scene. The sight of mountain mullet turn our thoughts to more familiar channels.

Leaving the trap, with Richard on guard, under a

magnificent banyan tree that grows in a field to the left of the pool, we started for the great fall whose voice we could hear in the distance. It is over a mile from the road, and is probably 150 feet in height and 175 in breadth. There is not one continuous sheet of water, but a myriad small cascades, feathery and brilliant, massed together, clustered, glancing at a hundred different angles, breaking into a thousand foam jets and playing with a broken rainbow that has got inextricably tangled among them. The habit that this eccentric stream has of throwing out terraces, ridges and dams is probably the cause of the bold promontory from which it falls. It has been built out inch by inch, and is still building, a living monument to nature's originality.

Back of Roaring River, in the Pedro district, is a Wesleyan high school at a place called York Castle. A village which goes either by the name of Claremount or Finger Post is on a side road, near the junction of the interior St. Ann's Bay road with that to Brownstown. Where the St. Ann's Bay road joins the Ocho Rios branch is situated the town of Moneague, a growing inland village with the usual complement of church chapels and government buildings.

# ST. ANN'S BAY.

WE drive into St. Ann's Bay, a town whose population approaches 2000 souls, late in the day, having finally succeeded in leaving Roaring River and its attractions behind us. At St. Ann's Bay reside several gentlemen who are prominent in matters influencing the prosperity of the country; and its business, especially in fruit exports is rapidly increasing, and it has, in common with Port Antonio and Montego Bay, direct and regular communication with the United States.

Not more than a mile away from the town was "Sevilla d'Oro" of the Spaniards: Golden Seville, founded by Don Juan de Esquivel, Jamaica's first governor. It seems almost incredible that in that early day, surrounded by what was then an unknown wilderness, there should have sprung up a city of which we read that the pavements of its cathedral extended two miles, that its theatre and palaces were splendid and its monastery world-famous.

We found a quaint inn kept by Mrs. Watson, on the main street. We were satisfied with our quarters and fare, and sought no further. It is a pleasantly situated house, with wide inn yard and many out-buildings. The recent epidemic of hotel building had seized Mrs. Watson, and she had in process of construction an addition to the house. Later the Honorable Michael Solomon, Custos of the parish, showed us, on an elevation back of the town, the proposed site for a new hotel which it is hoped that the growing needs of the place will demand.

A long interview with Mr. Solomon on the labor question showed that he stood as we had been led to expect, as a firm though not blind friend of the peasantry. While seriously admitting the faults of the race, which he (in common with the unknown author of the Blackwood's paper, from whom he quoted) held to be the faults of children, he spoke confidently of their advance.

He touched upon the fact that the plantations still put more money in circulation in the country than any other industry.

While we chatted, a woman came in and begged,—a buxom black woman of thirty-five years.

"I come to see de Custos, sah."

"Well?"

"I waanteh few shillin, sah."

"Have you no work?"

"No maastah—please Gawd."

"How many children have you?"

"I done got fifteen sah—dat all."

It was only a loan she required and the Custos, who is usually the guardian and father, especially of the poor of his parish, seemed to think she deserved it.

The house of Kerr & Co. dispatch a great deal of fruit and other produce to the United States, and

E. J. Wessels, a young American, has successfully established a business in this line that might well excite the envy of much older men.

From Mr. Wessels we have since obtained information upon fruit culture, and especially banana raising, which is as interesting as it is valuable, both to him who gathers facts for their own sake and to the practical business man who intends to use his knowledge as an investment, with his capital.

The good season for bananas (or rather the best season) is from March 1, to July 1, four months of good prices, which are followed by eight months of poorer.

Prices at the best are £15 per hundred bunches, and at the poorest £5 per hundred. When not grown three miles from the shipping place it is estimated that it costs £4 per one hundred bunches, to produce the fruit. The margin of profit is not so high as it seems, however, when we consider that a hurricane may occasionally mow down an entire crop; a remote contingency which, however, must be provided for.

At one time there was a great scarcity of inland transportation, to obviate which E. J. Wessels and Captain Baker exported great ox wagons to be used for this purpose. These the growers hire at a reasonable rate and thus are enabled to bring larger supplies to the market, to the benefit both of themselves and the shipping houses that handle the fruit.

The gross earning of an acre of bananas in one season is calculated to be at least £7 10s., which yields a tidy profit to the cultivator. Owing to the difference in soils the maturity of the fruit varies from nine to twelve months from the time of planting. It is most important to know this. One understanding the soil of any particular place, and being able to tell how long it will take the plants to bear will time his cultivation so as to take advantage of the higher prices of the spring market.

A good crop may be very much lowered in value by bad handling. Many growers crowd too many bunches into a dray in order to save transportation and so injure a great deal of fruit. Others cut and bruise the bunches by wrapping in "trash" (dry banana leaves, etc.) for transportation. The "trash" should be well watered and softened before binding the bunches.

At present, Mr. Wessels is making practically the first experiments that have been made towards the use of fertilizers. In many localities the natural productivity of the soil is such that it would laugh at such expedients.

As a final suggestion to any one who would embark in banana growing, test your soil first. Not all soil will grow bananas at a profit to the cultivator. The foregoing notes in the main apply as well to Port Antonio and all other fruit growing districts in Jamaica. The conditions are largely the same.

St. Ann's Bay possesses a handsome market, over the gate of which is a clock presented by Mr. Solomon, a few years ago. There are also three churches, Episcopal, Wesleyan and Baptist ; a court-house, constabulary station, goal and hospital.

On our way out of the town we pass the place where Golden Seville flourished, and where the great monastery and abbey stood in days of yore, but though we searched carefully over ground where they stood we found nothing, absolutely, of a city which owed allegiance to a Columbus, of a great religious establishment within whose walls the name of Peter Martyr was potent.

There is a village in St. Ann that goes by the uninviting name of Dry Harbor. That would suggest that it possesses prohibition principles, but the fact is that water is the scarce article. There is no fresh water at Dry Harbor except what is caught from the clouds in tanks, yet the place flourishes and trade increases. In the vicinity of this town it was that Columbus, upon his second voyage of discovery, landed, and took formal possession of Jamaica.

The neighborhood boasts a number of caverns. Not so brilliant and attractive as those of Kentucky or Virginia, but still very interesting to the visitor. Water percolating through white limestone does not offer the finest results in stalactites and stalagmites; still there is much to challenge the attention and waken imagination, as one proceeds a little way into the depths and views the intricate multitude of passages; dark labyrinths, that extend, for aught that is known to the contrary, for miles and miles under the hills. Doubtless there are lakes and lost rivers back there in the shadows. Many a sink hole drains the surface above and takes the water into some subterranean passage. Driving over the hills for long distances the

carriage wheels woke an echo under the road that
sounded like distant thunder. Probably all the western end of St. Ann is undermined, and perhaps ready,
in time of some great siesmatic disturbance, to subside
and change its topography, as did Port Royal and the
southeast once. The rushing of those subterraneous
streams "robs the pillars" of nature's mine, and a tremendous down-fall of the upper crust is the result.

Let us hope, for the sake of so gentle a saint, that
the parish named in her honor will not suffer such a
catastrophe.

Looking back from the road to Brownstown, the eye
beholds again those waving green fields and garden-like
plantations that help to make St. Ann so famous.
The road is built on the hillside and terraced well in
some places. One side often presents a bank of fern,
a bed of begonias or a cluster of broad-leaved " May
Poles," while on the other the wild fig and the Spanish
bayonet barely find foothold.

The soil changes to the deep red hue which affords
so brilliant a contrast with the outcropping limestone
that appears everywhere, and the vivid green of the
fresh fields and leaves. Huts by the wayside, framed
with bamboo, wattled with rush or cane and thatched
with dry banana leaves, have been daubed for preservation with the clay, which is so like ochre in its quality that it has been used in paint. As the clay in
various localities varies, so the colors of the houses do,
from chrome yellow to a deep maroon. These, with the
brilliant white of the walls, make a startling landscape.

Edgar Mayhew Bacon

Brownstown is one of the prettiest, trimmest little places imaginable. The tumble-down, squalid, sordid appearance that one or two towns on the coast possess, that look as though they had seen their best days, is utterly lacking in Brownstown. It has a clean, bright aspect about it, from the "Norman" New Market (named after Sir Henry W. Norman), at the foot of

the hill, to the Tabernacle built by the evangelist, Dr. Johnson. Being a hill town, there is little level walking possible in Brownstown, but the air is so good that under its influence the hills forget to be fatiguing. Dr. Johnson is the presiding genius of the place: preacher, physician, builder, photographer and adviser to half the population. We are indebted to his excellent photographs for some of the illustrations of this book. The people find his tanks almost miraculously filled in time of drought, and the supply free to all who ask, while others are selling water.

We will not say that he is beloved by all, because he is too forcible a character for that ; but a great many men would be proud of his personal following.

Brownstown is the commercial clearing house for the hill region of the interior. One of the leading merchants, Mr. Levy, has a number of shops, carrying a stock that might suggest big possibilities to American manufacturers and wholesale merchants.

There is in Mr. Levy's employ an old man whose general usefulness is acknowledged and whose eccentric views are evidenced by a coffin which he had made for himself several years ago and which, having it hanging where he can feast his eyes on it every day, he enjoys in anticipation. For thirteen years it has waited for him.

Besides the buildings noticed there is in Brownstown a good court house and police station, a very pretty Episcopal church, a Baptist place of worship and a remarkably comfortable, clean, and inviting lodging, kept by one Mistress Delisser.

Stewart Town, on the Brownstown and Falmouth road, is principally noteworthy for having in its immediate neighborhood one of those sudden resurrections that seem to be the peculiar characteristic of the rivers of this part of the island.

The Rio Bueno, which was probably the Sink River in the southern part of Trelawney, bursts suddenly from the ground and escapes to the sea, and is the boundary line between the parishes of St. Ann and Trelawney. It empties near the town and bay of Rio Bueno, once an important shipping port.

# FALMOUTH AND VICINAGE.

THE Martha Brae is a river that is uncertain whether it should run like Aph, the sacred river described in the poet's dream, "through caverns, measureless by man," or whether its best course is a frank above-ground one. Doubtless its first appearance was in the Mouth River, that drops quietly into its burrow after awhile and defies pursuit or detection. From the place where it comes to daylight again the Martha Brae runs a sinuous course to a town once known as Melilla, near Falmouth. That was a place that the Spaniards built when they first landed in Jamaica and afterwards abandoned. It is now known by the name of Martha Brae.

All the world knows that the powerful loadstone which drew most Spaniards to the New World was gold; and Martha Brae has its golden legend.

In Sir Augustus J. Adderley's sketch of Jamaica appears this paragraph:

" In the ' Notes to Thomas Burton's Diary ' (Clarendon State Papers), will be found this curious remark: —The secret golden mine which hath not yet been opened by the King of Spain or by any other is four miles from Mestan towards the east. It is near the way towards Mellila. The earth is black—rivulets

discover the source of the mine. The name Sevilla
has survived, and is now the parish of St. Ann, and a
French author tells us, writing in 1660, that the town
of Ovistan, built by the Spaniards, is not far from a
bay or river in which the Bluefields River disgorges
itself. Bridges has also traced the spot where once
stood the town of *Millila* on the banks of the *Martha
Brae River*, so that we have *Sevilla* in *St. Ann's
Ovistan* in *Bluefield Bay*, and *Millila* on the banks of
the *Martha Brae*. The so-called secret gold mine
must therefore be somewhere in this region, if it exists
at all. The story goes that the Spanish Governor,
Don Pedro d'Esquimel, extracted the secret of its
existence from an unfortunate Indian chief by the
usual means of torture. The wretched man had
appeared before his Excellency only too magnifi-
cently decorated with golden ornaments, and thereby
awakened his appetite for possessing a knowledge as
to the spot whence the chief obtained such riches.
"'Were I to search for the famous secret mine,' says
Bridges, 'I should look for it on the Maxfield estate
and in the neighborhood of Trelawney.'"

Falmouth has a population of between three and
four thousand souls. Its pride is in its public buildings, its water supply and its history. Its prison
buildings are unusually fine and well kept, though
fortunately not unusually full. The court-house in
the centre of the town, is one of the best on the
island, contains offices for nearly all the parochial
boards and officers, and has a ball-room of delightful
size, upon the walls of which are two full length por-

traits of former governors of Jamaica. The churches of Falmouth are fine edifices, the Baptist chapel being especially so. There are military barracks capable of containing 700 men, but at present a force of about thirteen " constabs," occupy (?) them. To get to the almshouse and prison the shortest path is across a causeway which divides as fine a musquito nest as any place can boast. The water supply drawn from the Martha Brae is stored in a large tank in the centre of the town. Thence it is distributed by pipes, or drawn by the people, who come with their various vessels to draw.

Though still important, there is a general complaint that the business of Falmouth has fallen off of late years. This is supposed by some to be due to the great rock in the harbor's mouth, which is a very serious impediment to large steamers. The old sailing craft being smaller did not find the same difficulty in escaping it. Until that rock is blown up and the harbor made safe, Falmouth must expect to lose by it.

Some of the business men and land-owners of the vicinity complain of a scarcity of labor. Everywhere, where there is such a shortage, the sugar planter feels it first. The estates have less hold on the laboring people than any other class of employment. But the majority of people in Falmouth are disposed to be very serious over the scarcity of employment which is a more vital matter. As yet the district produces little or no fruit.

# MONTEGO BAY.

ONE of the best places in Montego Bay from which to view the town is the large house kept as an inn by Mr. Payne. There are other lodgings in the town, but Mr. Payne's is one of the pleasantest and best on the island. Its windows overlook the harbor and from its vantage above the heat or the night dampness of the lower land, we are inclined to believe that beauty is, after all, the thing most to be valued in life (after a good dinner), and we have secured them both.

Under the guidance of Mr. Corinaldi, the United States Consular agent, we visited the court-house and the church, admiring in the latter some good statuary, studying the mortuary tablets with which the floor is paved, and mounting under the fine-toned bell to the tower from which the amphitheatre of hills can be viewed that surrounds the town.

In that church we were shown the memorial erected years ago to a good and beautiful woman—good, according to the inscription on the marble, and beautiful by tradition. Yet this gentle saint was pointed out to all comers for many years as an utterly depraved character, a murderess, in fine, whose hands had been dyed with the blood of her own husband.

Time is a great indicator. One day some one discovered records which clearly proved that not this
woman but another of the same name had committed
the deed for which for years this marble has blushed.
In spite of the contradiction (which may not yet have
reached the land where avenging ghosts have their
habitat), there was when we called a tiny spot of *fresh*
blood on the stone at the base of the monument.

While we are loitering within the sacred walls there
is much to be seen in the busy town. Montego Bay
has a population of four or five thousand people. It
is commonly supposed to rank next to Kingston in
commercial importance. Messrs. J. E. Kerr & Co.,
the owners of a steamship line running to the United
States, have their headquarters here, and the Atlas

Company's vessels and others visit its harbor regularly, having their agents in the place.

The name originally was Manteca Bay, " Butter Bay," or, rather, " Lard Bay," for it was in the latter commodity that the Spaniards dealt most extensively. A general shipping business in the staples of the country has been largely augmented by the increasing fruit trade, and it is said that many properties in the neighborhood which have been considered almost valueless, have become profitable as fruit lands. Besides this the people are more generally employed and are more contented. The chief buildings in the town are the court-house, the Episcopal Church and Trinity Chapel, the chapels belonging to the Wesleyan, the Baptist and the United Presbyterian denominations, the custom house and the old barracks.

We have referred to J. E. Kerr & Co., as having their headquarters in Montego Bay. That firm deserves more than cursory mention, since its enterprise has largely been the means not only of keeping Montego from the condition of decadence into which so many Jamaican seaports have fallen, but has actually built up and improved the commerce of the place to such an extent that now the outlook for the future is even brilliant. Added to this, their steamers are known to be among the fastest of those steaming among the West India Islands. The parish of St. James generally feels this impetus given to trade. Like most of the north ports Montego Bay's commerce is largely with the United States. Montego Bay has, within a few years, been greatly improved from a sanitary standpoint by the filling in of a swamp in the neighborhood. Of all the towns of the North and West this one has shown first interest in the Kingston fair. There is here a fine market, and the usual complement of public buildings.

# IN HANOVER AND WESTMORELAND.

"HAVE you been to Lucea? Do you not think it one of the finest towns in Jamaica?" is a question which we have had frequently to answer. Lucea is a beautifully situated little town of nearly two thousand inhabitants. It is near the northwestern point of the island, where the coast begins to slope to the southwest. Its harbor is a deep, almost circular basin, much narrower at the entrance than inside, so that in approaching from the east we are much nearer the town when we first come in sight of it, than when a half mile further advanced in the road. The houses, business buildings and court-house are on low ground facing the east and north, while above them on the hills are pleasant residences, and picturesque grounds. Here also is a fine old church, and beyond it but a little distance, the clean and well-kept prison. The fort (Charlotte) at the entrance to the harbor has been converted into a constabulary station. Besides the Anglican Church Lucea has a Kirk and Baptist chapel below the hill. The climate of Lucea is recommended as being exceedingly good, and statistics show the place to be a healthful one. We were told of one or two lodging-houses in Lucea but chose one that had been specially recommended to us, kept by

Mistress Vosper. We found the lodgings fair and the board better. The only drawback was a drain or canal which a continued drought had left in a stagnant condition, running next to the house and filling the air with an odor that was not perfume.

The Rev. Mr. Davis, rector of the Episcopal Church, is our authority for the statement that in Hanover exists the same scarcity of laborers that is complained of in other parts of Jamaica and for which there is suggested no remedy but immigration. At the time of our visit, early in July, we found a very slight awakening interest in the coming industrial exhibition at Kingston. Wherever such an interest has ripened it has been by active personal endeavor on the part of representative men, and largely through the establishment of local preliminary fairs.

We left Lucea on the morning of the 4th of July. Such of our readers as have been accustomed to regard the Stars and Stripes with reverence and affection, can understand how we searched (vainly) for a flag that should bear the familiar arrangement of red, white and blue. Nor were our efforts to secure a pack of " fire crackers " more successful. As we drove up into the hill country we tried to sound Richard's ignorance on the subject of American Institutions. He had heard that the Fourth was some kind of a thanksgiving day, but he did not know who George Washington was. Neither could he learn to whistle The Star Spangled Banner. We did our best, but it was a lonesome festival.

Our anticipation of finding in the interior of Hanover the Switzerland of Jamaica was doomed to disappointment. The " Dolphin's Head " Peak that loomed up to the left of our road was certainly charming, and would be an ornament to any landscape, but it was not Alpine in its proportions nor in its contour, being two thousand feet high. Neither did the little cabins by the wayside resemble the Swiss *chalets* in anything but size. The vegetation seemed rather less tropical than that of the eastern end of the island. But it is a beautiful road, offering numberless surprises to the traveller and presenting at almost every turn some new delight in scenery.

Our objective point was Savana la Mar, to which three roads converge. We were on the central one. The more westerly takes in the village of Green Island, and unites with ours at the border of Westmoreland. The eastern road is a more direct course from Montego Bay. Some of the most valuable grazing pens on the island are situated in this part of it. Here is Knockalva, whose Hereford cattle are famous, and over whose thousands of acres of pasture land noble herds are bred and reared. Hanover has more than four times the acreage in pasture that it has in cultivation, and both together do not half equal the number of acres in woods and ruinate. The logic of such a fact leaves no room for argument on the question of under-population.

We have passed many gardens where the famous "Lucea yam," which has been so valuable a product for export to Colon, is cultivated. To the west lie the hills between the Fish River and Negril's, beyond which are the marshes of Long Bay, which the alligators are said to haunt. We have passed into Westmoreland, and begin the descent through a sugar-growing country towards the plain of Savana la Mar, which, like Kingston, is built on low ground. The two greatest disasters of Jamaican history did not occur in the mountains. The earthquake which swallowed Port Royal was almost equalled in horror half a century later by the tidal wave which swept Savana la Mar and all it contained of human life and property from the face of the earth. We are struck (as we drive in through the rain which met us on the southern hillslope) with the appearance of the single, broad, central street which runs inland at right angles to the shore. The reason for this position is apparent; the old city, the city of the tidal wave, having been built along the shore, at present the court-house, a half-mile from the wharves, marks the limit of that devastation.

Savana la Mar is an important seaport with a population of between two and three thousand souls. It contains several large business houses, and has postal communication every day with other parts of the island. There are Episcopal, Wesleyan and Baptist Churches; the building of the Wesleyan being new, and all the places of worship commodious and services well attended. The water supply is excellent. There is good medical attendance to be had in the place, we are told, though, fortunately, we have had no occasion to prove the accuracy of the statement. Nearly opposite the constabulary station we find lodging at a very pleasant little hotel, where clean beds and good food are furnished at a cost of 10*s*. per day ; there are also several other lodgings in the town. Truth compels the admission that Savana la Mar is not a beautiful place, either for situation or architecture, as compared with other Jamaica towns.

The town of Blue Fields, about eight miles distant, is worth visiting, as the site of Oristan, of the Spaniards, and also as the home of Gosse, the great naturalist. The cost of conveyance thither is 1*s*. 6*d*. per mile. Other places of. interest in the neighborhood are Roaring and Amity caves. At a point called Paul's island some coolies have started the cultivation of rice, which bids fair to be a remunerative venture. Westmoreland produces besides this sugar and rum, coffee, spice, logwood and some fruit.

# ST. ELIZABETH—THE SANTA CRUZ MOUNTAINS.

LEAVING Savana la Mar our road keeps near the coast through the southeastern part of Westmoreland and into St. Elizabeth Parish as far as the town of Black River. Beyond Bluefields we pass through the country of the Surinamese settlement. Crossing numerous streams, and journeying through a fertile country with many glimpses of distant mountains or of the nearer ocean, we approached one of the largest and most important of Jamaica's navigable streams. The Black River is about two chains in width near its mouth, where the town is built. It is there spanned by an iron bridge, that was erected several years ago at a cost of £1700. For thirty or forty miles of its sinuous course, as it winds over the lowlands and between the uncertain margin of an extensive morass, this river floats little vessels of eight or ten tons burden, which are used as lighters for logwood, thus shipped from the interior. There is in the town of Black River one long central street, not attractive to the visitor. Upon this street is the hotel. We looked out from the hotel at the "Daisy" lodgings opposite and wished that we had gone there : but perhaps, after all, we were as well off a. we would have been elsewhere in the place. Of the "cleanness, civility and good attendance" promised on the neatly printed card of the Waterloo Hotel, we found an abundance of the second item and a moderate allowance of the others. Besides the "Daisy," there are the "Britannia" and private lodgings kept by Mrs. Allen. The cost of lodging and board is from 10s. to 12s. per day, which is the usual rate in other parts of Jamaica. The rule seems to be that at the best places one is charged 10s. and at inferior ones 12s.

From Black River the road winds by the border of a wide and extensive morass, and is sometimes submerged for long distances. It crosses the Y. S. River, upon which at the distance of a few miles only, are the rather famous falls; and after passing and repassing the loops and twistings of the Black River, begins to mount into the higher land towards the Santa Cruz Mountains.

The Santa Cruz Mountains of the Parish of St. Elizabeth are famous as a health resort, though not so frequently visited as the region of St. Thomas ye East because of their less accessible location. Here in the hottest weather the nights are surprisingly cool and the wind at times blows almost a gale, while with two blankets we are quite comfortable in July. Yet the elevation is but little over two thousand feet, and the most desirable points can be reached from Black River by a three hours drive.

The village at the foot of the mountains is Lacovia. Here the roads branch, one going by the way of Santa Cruz Square towards Mandeville, and the other into the Nassau Mountains. At Santa Cruz Square, a cool and pleasant village, there are lodgings to be obtained at a cottage near the road, and simple good fare is offered. Our Mecca is Malvern Hill, to reach which we must give the horses one of the hardest day's pulls they have yet had: but if horse kind enjoy those luxuries of life, the influences of air and surroundings, as human kind do, they were repaid for their work.

Malvern Hill overlooks sixty or eighty miles of coast. One can stand on the breezy piazza of Mrs. Lawrence's pleasant home, where well accredited visitors are entertained at moderate price, and overlook the region from Savana la Mar to the Pedro Plains. It is at two thousand three hundred feet elevation, but its average temperature is from 70° to 76° in daytime and 60° to 66° at night. The wind blows strongly always. At night, when one is safely sheltered and blanketed in the well-built house, its surf-like roar is pleasant. The inland view is almost equal to that of the coast, embracing as it does a parallel ridge. Malvern is eight miles from Santa Cruz Square, forty miles from Porus, and ninety from Kingston. The shortest way to reach this point from Kingston is by coastal steamer to Black River, fourteen miles away, and then by carriage or in the saddle from the latter place. A double-seated "omnibus" from Black River to Malvern Hill will cost 36s. sterling ; a single buggy about 10s. less. Saddle ponies cost less.

There is not space to dwell upon the beauty of the

lover's leap at the seaward end of the Santa Cruz
Mountains, or of the great Pedro Bluff, that juts out
boldly into the sea but a little way to the westward.
We must not linger along the intricacies of roads and

cross roads in which we were lost in trying a short
cut to the main highway, nor rest after our tramp
up "Seven Mile Hill," which we take partly because
it is a pleasant day for walking and our legs are
cramped with weeks of driving, and partly because " a
merciful man is merciful to his beast." We solace
ourselves with a " Ripley " pineapple and push on
towards Mandeville.

For many years the hill-region or plateau of upper
Manchester has had a widespread reputation as a
sanitarium. The Rev. H. Walder, a Moravian missionary who has had a wide opportunity to form an opinion, considers this region quite able to bear comparison
with the healthiest regions in England or on the continent of Europe. He compares it favorably with the
Canton Zurich, Switzerland, quoting his experience
with the small congregation at Mizpah. An average

membership of 473 yielded but 60 deaths in twelve years, and during that same period there were 241 baptisms of infants. This would seem to indicate a birth-rate four times as great as the death-rate. Of the 60 whose deaths were recorded, one was one hundred years old, seven between eighty and ninety, seventeen had passed seventy years, six were over sixty, four died from accidents and ten as infants. Thus, allowing for accidents and the deaths of infants, over 67 per cent. lived to be over sixty and over 54 per cent. passed man's allotted three-score years and ten. Truly a most wonderful record ; all the more wonderful when we consider Mr. Walder's closing words : " Taking furthermore into consideration the innumerable wettings which the generality of the people are exposed to, without even being able to put on a dry suit, and, in addition, the very great defects in the homes of the peasantry, in a sanitary point of view, it cannot be denied that the above facts speak loudly in praise of the climate in these districts."

# THE HOME STRETCH.

WE drove into Mandeville by a road that showed many signs of contented unthrift. There were miles of fertile country, an exquisite land to travel through, but inadequate cultivation on every side.

Finally the houses became more numerous and much better than those we had been accustomed to seeing along the road, and we drove into a trim open square. In the centre of the square, with an acre or two of grassy common surrounding it, stands the court-house. At its right is a row of shops and other build-ings. At the side on which we enter, facing the court-house, is the church and several dwellings. On the left and on the rear are other dwellings, shops and offices. In that direction are also the hotel and one or two lodgings. Not far away is the house kept by Miss Roy, where Mr. Anthony Froude sojourned during his short stay here. His praise of her coffee has made it famous. We went however down the little hill to the hotel, Brooks', where we had no cause to complain either of lodging, fare or attendance.

Mandeville, on its table land, is as charming a place to the eye as it is beneficial to the sense. Its general aspect is one of extreme neatness and prosperity,—and this in spite of the clay mud which we found omnipresent, mud that was like paste, which clung in great yellow clogs to the shoes of whoever walked abroad. But who cares for mud in a climate like that?

There are many points of similarity between most Jamaican towns, usually in the arrangement of streets and dwellings, and almost invariably in some suggestion of a historic flavor. But Mandeville, like Brownstown, seems new. It is not aggressive and unpleasant newness, but a neat and engaging condition of youth; youth with plenty of color and tone.

Our time and space are unfortunately limited: an unfortunate fact, for we are in a country where a limit is the thing least to be desired; there one should be allowed by Providence to roam on forever to study and to enjoy.

We are in our places again, in the trap that has carried us well for over four hundred miles (may the Wagonette Company's shade never grow less), and Richard,—whose arrivals and departures are a matter of gaping wonder to the black people by the way,— takes the road with much whip-snapping. Out of sight of the houses of Mandeville our pace slackens into a steady jog, and we draw into a section with which our railway trips have made us familiar. This is the region of the Mandeville orange, well known in Northern markets from its size and flavor. The road is a more level one, with interesting features, but nothing as startling as we found at points along the Northern side. All this part of the country is the best horse raising land on the island.

The little town of Porus, the present lower railway terminus, is reached and passed. There is nothing here to entice us to linger now, though the natural beauty of the surroundings of the busy and populous little place induce another visit. At May Pen we cross its handsome three truss iron bridge and wonder at the want of precaution that will permit the railway trains and a constant line of vehicles and foot passengers to cross in common on the same structure.

Thence to Old Harbor; or rather to the real town and railway station, two and a half miles from the original village of cabins and small houses which bears the name. At Spanish Town, from Mrs. Wilson's very pleasant lodgings, another stroll down the Rio Cobre renewed all the fascination of that beautiful stream. Its clear, ever varying water, shadowed by magnificent trees, is a perpetual delight.

The road to Kingston is bright with patches of cultivation and more and more filled with people as one approaches the metropolis. The great cotton tree known as Tom Cringle's tree spreads its giant arms over the road and challenges attention; but Jamaica has shown us, on its mountains and by-roads many a finer trunk and broader, loftier expanse of foliage than this. We have seen buttressed roots that would have walled a dozen cabins and a spread of foliage fit to shade a herd of a hundred grazing cattle.

Finally we approach Half-way Tree, turn out towards Constant Spring Hotel and ere long have washed off the dust of travel and are answering questions.

"What of the island? does it realize expectations?"

"Yes, and exceeds it."

"And how about the living?"

The answer is given to the reader in the preceding chapters. He who follows our itinerary will neither lack by the way nor suffer on his return.

A number of points, not included in the accounts of our various trips, have more or less of interest attaching to them, in most cases from some one or two peculiarities of natural position or advantage.

Among these are Milk River, noted for its bath of curative water, on the coast of Clarendon; Morant Point, with its lighthouse, in St. Thomas ye East: and several interior points out of the ordinary lines of travel.

The unsettled Cockpit Country and Black Lands in St. James and Trelawney Parishes are full of interest and well worth exploration. The mystery of lost rivers, of cavernous hills, of wild, weird forest tangles, of mysterious sink holes and sudden cliffs, lure one to adventure. North of the Surinamese settlement, which we passed in Westmoreland, where the Dutch Colonists set an example of industry to the other settlers two hundred years ago, there is a fertile country dotted with villages and estates.

Bull Head, Chapelton and the district of Vere in Clarendon and Mount Diabolo, above St. Thomas-in-the-Vale, all invite the traveller. We have not climbed the John Crow Hills together to wonder at the recently discovered lake or morass at the summit of the ridge. We feel that we have neglected Buff Bay and have not done justice to Luidas Vale. We hope to be forgiven for our delinquencies.

The Hotels law of 1890 insures the presence in many of the pleasantest parts of the island, of modern hotels largely managed according to the methods usual in the United States. In other localities very fair lodging or boarding houses may be found, and in all of these the degree of comfort desired can nearly always be had for the asking, for the Jamaica landlord—or landlady more often—is full of the desire to make life comfortable for guests.

# TRANSPORTATION AND COMMUNICA-
# TION.

IN writing of the pleasures, attractions and notable features of any country, he who does not also inform his reader how these advantages may be shared, is like one who exhibits a treasure chest and afterwards returns the key to his own pocket. It is, at best, an unsatisfactory performance.

It is within the scope of the present work to publish minute information regarding the means provided for travel and communication between Jamaica and the American Continent.

We found, when contemplating a visit to the West Indies several years ago, a scarcity of information, which, considering the distance, was almost prohibitory. At last, when nearly ready to abandon the quest, we received a copy of the Atlas hand-book, then just published, which removed much of the difficulty.

Perhaps it is because of that association and the pleasant experience of a first voyage that we begin our chapter on the means of communication with an account of the Atlas line. There is also another reason, however, and a better, which is, that the oldest and best line running from New York to Kingston is the one which flies the familiar blue flag with its white-centered red cross.

The company employs a fleet of twelve large iron steamships, of which several of the best carry passengers and freight on the Jamaica route. The *Adirondack*, the last and largest of the fleet, with its comfortable arrangements and latest appliances of science ; the *Alvo*, stanch and steady as a church, ample as to state-rooms and satisfactory *en cuisine ;* the *Athos*, *Ailser*, and *Alene*, all able boats and comfortable carriers, are constantly on the way to and from the island. All of these vessels are well officered, and offer every inducement in the way of comfort and good service to the traveller. The advantage of having the saloon on deck has been often gratefully commented on by passengers who are subject to the discomfort of *mal de mer*, and would find it as impossible to dine in an apartment below deck as it is easy to do so in comfort above.

The trip to Jamaica takes five and a-half days ; half of that time being spent on pleasant sub-tropical waters, through the deep, constant blue of the open ocean, or among the turquoise and emerald hues of the Bahama waters. One embarks at Pier No. 55, (New) North River, foot of Twenty-Fifth Street, whence an Atlas boat leaves every Saturday for Jamaica. In returning, the vessels leave Kingston on alternate Thursdays. By regular arrangement for round trip or excursion tickets, either the direct travel to and from the island may be enjoyed, or the voyage extended to include the circumnavigation of Jamaica, or a run over to Haïti, or even the interesting ports of Central America may be visited.

Lately a regular steamer, the *Arden*, has been put on the Haïtian route, stopping at various ports and running to Port au Prince. The *Adula* makes the coastal trip, stopping at all the principal insular ports. This is not only a delightful and instructive addition to the itinerancy of visitors and vacationists, but one of the best facilities afforded to business travellers, and an outing often. enjoyed by Jamaicans. This coast-wise steamer sails from Kingston every ten days, alternately to the eastward and westward. Passengers may also connect, by excursion routes, with all the best European lines.

But it is not only to the travelling public that the Atlas line affords peculiar advantages. Shippers find the convenience of transhipment to the coasting steamer in Jamaica, or enjoy the facility of having the port of destination on through bills of lading to Europe changed, or goods stopped and delivered in New York, if the company receive sufficient notice. Cargo may also be transhipped at Kingston to the Royal Mail packets, through bills of lading being issued by the Atlas Co., in conjunction with the Royal Mail Co.

The steamers of the Honduras and Central American S. S. Co., leave New York every three weeks, making stop first at Port Antonio and then at Kingston, *en route* for Central America. They are about 1,700 tons register and are fitted to carry passengers.

We have spoken in another connection of the steamers of Messrs. E. J. Wessels & Co's line. Besides being a buyer and shipper of island produce this company has provided excellent passenger accom-

modations on its fast steamers. These are, like those
of the Atlas boats, above the upper deck. The vessels
sail from the north side of Jamaica (St. Ann's Bay)
to New York, twice a week.

The Royal Mail Steam Packet Co's. vessels carry
freight, mail and passengers to and from Southamp-
ton, via Barbadoes. The Clyde Line to Glasgow
affords a monthly steamer. Its new steamship, the
"Strathyre," is especially well noticed.

J. E. Kerr & Co. are running good steamers to the
United States from the North Side, as is also the
Boston Fruit Company, whose Jamaican headquarters
at Port Antonio we have already described.

There is a strong probability of the immediate
establishment of a line of small but fast and sea-
worthy steamers to run between Galveston, Texas, and
Jamaica. The line was organized last summer with
the purpose of visiting Central and South American
ports, taking Kingston, Jamaica, into the scheme.

A still more recent mode of conveyance, especially
desirable for those who wish to visit the Antilles, yet
avoid the discomforts of a long sea voyage, is that
included in the Plant system. This provides steam
communication with our island by way of Tampa,
Florida. This will enable the traveller to economize
both time, money and strength, if he is one of the
unfortunate ones to whom a period of five or six days
on the ocean is less than a boon. The first steamer of
this service was announced for October, 1890. It is
impossible to give here further details. During the
winter of 1890-91 the Tampa steamers will run fort-

nightly, or possibly weekly, and their continuance beyond that season will depend entirely upon the success of the experiment.

We have left to the last one of the most important lines now engaged in the transportation of passengers and freight to Jamaica. This is the Anchor Line, whose steamers are now plying between New York and Jamaica.

On the third day of November, 1888, the Anchor Line of Steamships, whose transatlantic vessels are so well known to the world of travelling people, made its first venture in this field. The Steamship *Dorian*, built for the Mediterranean fruit trade, was put on the Jamaica route and sent to Kingston, whence she visited all the principal island ports and returned straight from Jamaica to New York. The *Dorian* was shortly followed by the *Tyrian*, also designed for the Mediterranean trade, and these vessels have continued to ply regularly between the island and New York.

Their trips are made fortnightly. They are mail carriers, and the first mentioned, the *Dorian*, is fitted with passenger accommodations, although this line does not make a point of passenger service.

The present accommodation of the *Dorian* is for about twelve first-class passengers. Besides the vessels described, the steamers *Sidonian* and *Acadia* occasionally make the Jamaica trip, but not regularly.

The principal business of the Anchor Line in Jamaica at present is the shipment of fruit and other cargo. As stated, its route is a direct one to and from Jamaica. Stopping at Kingston first the steamers land

and take freight at Port Morant, Black River, Savana la Mar, Montego Bay and Lucea, St. Ann's Bay, Falmouth, Port Maria, etc., and thence at once to New York. To shippers the company offers the inducement of through bills of lading, via New York, to principal European ports, transhipping in their own vessels.

It is proposed to add shortly passenger steamers, especially designed for the work, to those already in use.

We should add a few more explicit words regarding inland travel in the island, giving a *resumé* of information already detailed.

Besides the railway, there is a mail coach which communicates with all the principal towns and affords means of travel usually three times a week, at rates which are less than those of carriage hire. From the Wagonette Company, in Kingston, one of whose conveyances we used in our trip 'round the island, can be procured good horses, wagonettes and buggies. As to its drivers,—you have already made Richard Davis's acquaintance. Its charges are not materially different from those of Bolton and other livery stable keepers. £1 per day is the usual rate for a double trap retained for any length of time, though when taken for a single day or short trip the rates are of course much higher. In Kingston the average price is 6*d.* a mile : for a day 30*s.* to 35*s.* for busses. In Hanover, Clarendon, St. Catherine, Portland or St. Mary there are no livery stable keepers. Good saddle ponies are hard to get.

At Savana la Mar J. B. Jones charges at the rate

of 1*s*. 6*d*. per mile. Fair conveyances can be obtained in Porus, Mandeville, Falmouth, Brownstown, St. Anns, etc. At Gordontown, Bolton keeps good saddle horses for mountain riding.

The tram cars run from Kingston to Halfway Tree and Constant Spring carrying passengers at a charge of 2*d*. a mile.

# POSTAL AND TELEGRAPH FACILITIES.

In sending letters, postal cards, etc., from point to point in the island of Jamaica, or abroad, the rates of postage charged are as follows:

## LOCAL.

| | |
|---|---|
| Letters: for each half ounce, . | One penny. |
| Post cards: (inland only), . | Halfpenny. |
| Post cards: reply prepaid, . | One penny. |
| Newspapers, . . . | |
| Current prices, . . . | Halfpenny. |
| Book packets (for each two ounces), | |
| Parcels (for each two ounces) . | One penny. |
| Registration fee, . . . | Fourpence. |

## FOREIGN LETTERS.

Not exceeding half an ounce, twopence halfpenny. Not exceeding one ounce fivepence.

For each additional ounce, twopence halfpenny.

A book packet may not exceed 3 lbs. in weight. Packets containing value must be registered, failing which they are liable to a surcharge equal to double the registration fee. Parcels containing articles valued at above £50 will not be accepted for transmission by post; otherwise parcels under 11 lbs. may be sent to the United Kingdom or the United States of America, and under 7 lbs. *via* Great Britain to all countries included in the postal union.

Money orders are subject to the following charges:

| WHERE PAYABLE. | COMMISSION ON AN ORDER. | | | | | | | |
| --- | --- | --- | --- | --- | --- | --- | --- | --- |
| | Not exceeding £2. | | Above £2 and not exceeding £5. | | Above £5 and not exceeding £7. | | Above £7 and not exceeding £10. | |
| | s. | d. | s. | d. | s. | d. | s. | d. |
| In the United Kingdom and for through Money Orders.................... | . | 9 | 1 | 6 | 2 | 3 | 3 | . |
| In the United States of America.......... | . | 9 | 1 | 6 | 2 | 3 | 3 | . |
| In Canada............................ | 1 | . | 2 | . | 3 | . | 4 | . |
| In British Guiana and Barbadoes and Leeward Islands..................... | . | 6 | 1 | . | 1 | 6 | 2 | . |

On the last day for issuing Money Orders by each mail an extra commission of *sixpence* on each order will be charged.

With all first-class foreign powers, with the dependencies of Great Britain and with the United States Money Order exchange has been authorized.

The telegraph system now in operation on the island was established in 1879. It embraces all principal towns and villages at a uniform charge of 1s. for the first twenty words and 3d. for every additional five words. The telegraph is a branch of the

P. O. department. Jamaica, in common with Barbadoes, Trinidad, the Leeward islands, the Windward islands and Demerara pays an annual subsidy to the West India and Panama Telegraph Company (cable) according to the terms of an agreement made in 1879, by which the cable company furnishes daily a summary of the world's news and prices current for the general benefit of the public. The tariffs for messages to North America and Europe, *via* Havana are from 4*s*. 10*d*. to 6*s*. 9*d*. per word ; the lowest rate being to Key West, Fla. Communication by this channel is open also to the other West India islands.

# PUBLIC WORKS.

UNDER the head of public works are included main roads and bridges, buildings belonging to the government, lands, canals, gardens and light-houses.

For the first, few countries in the Western Hemisphere can boast as good roads as Jamaica can. They were almost the first object of legislative care, as early as 1681. Nearly thirty years ago they were removed from parochial control and assumed by the government. The main roads are the coast road, now almost continuous, which begins at Kingston as the Windward road, leaves the shore in a few places, especially in the east and southeast, but follows generally the outline of the island; the new inland road from Kingston to Annotto Bay; the Spanish Town and St. Ann's Bay road, with its branches to Port Maria, Ocho Rios and Falmouth; the upper road in Manchester and St. Elizabeth, and the short cut from Lucea to Savana la Mar. There is continual enlarging, widening and repairing of highways, repairing and building culverts, and retaining walls and bridges; yet it is doubtful if the cost per annum for maintenance, repairs and new work throughout the whole island will average over £50 per mile. The best road that we found in Jamaica was that between Morant and a point beyond Bath. The total absence of frost is a saving to the public works department. But occasionally a heavy storm, such as that which occurred in 1885, does incalculable damage in a few hours.

The finest bridge on the island is that over the Dry River. It was built in 1874 at a cost of £16,901. This structure is of iron, and consists of three spans, the centre one being 150 feet in the clear. Its girders are 40 feet above the river bed, which is usually dry, though the water sometimes reaches a height of 35 feet in a flood. The Annotto Bay, Black River, Flint River, Ocho Rios, Falmouth, Montego Bay and old Harbor bridges are among the most important of these works. There are also new bridges in St. Mary, Portland and St. Thomas, some of them still in process of construction.

The work on these roads and bridges is done in the most thorough and business-like manner.

Public buildings comprise churches, certain chapels, hospitals, and schools, and all courts and government offices, prisons, reformatories, police stations, asylums, etc., besides the Kingston Library, the Printing Establishment, and some few others.

The Church of England receives governmental support throughout the island. Every town of any importance has its court-house, jail or police station, and hospital. These buildings are substantially built, generally of stone or iron. The prisons are worth visiting. Their management is usually praiseworthy. Everything is orderly, clean and comfortable as a rule.

Indeed the care bestowed upon prisoners, inmates and patients in the various classes of institutions mentioned, the enforcement of sanitary regulations and the quality of food provided all give· evidence of the most efficient departmental care.

The most important canal work ever attempted in the country is that in St. Catherine, the Rio Cobre irrigation system, which we have spoken of in another chapter. This work was completed in 1876. The head works are situated four and a half miles above Spanish Town. They consist of a weir for raising the level of the river and sluices for admitting water into the canal. The crest of the weir is 30 feet above the bed of the river, and it has a clear overfall of 287 feet. The average flow is 45,000 cubic yards per hour. The canal required bridges, aqueducts, walls and culverts in its construction. An iron pipe (36 inch) syphon, 6 chains long, carries the water beneath the Rio Cobre at one place. It has a total length of nearly forty miles.

Of gardens Jamaica has a rich share. The Castleton garden we have seen on the road from Constant Spring and Stony Hill to Annotto Bay. That of Bath has been practically abandoned, Hope taking its place in a large measure. The difficulty with the Bath garden was its complete inundation at certain times. The cinchona plantations comprise 143 acres in cinchona, and about seven in tea and nurseries of timber and shade trees, which are distributed for reforesting. These plantations are twenty miles or more from Kingston, on the slopes of the Blue Mountains in St.

Andrew's. The Hope nurseries are on the Gordon-town road, about five miles from Kingston. New varieties of fruit and shade trees, timber, cane, etc., are raised here. The annual mean temperature is 77.8° Far., and the rainfall averages 50.19 inch per annum.

The King's House gardens and grounds, the Palisa-does Plantations and the Kingston Parade Garden are respectively devoted to economic and fruit trees, and rare flowering plants, to cocoanuts, fibre plants, etc.; and to shade and ornamental growths. It is largely through these experimental grounds and cultivations, often kept up at great cost, that Jamaica has become the garden spot it is to-day. Since 1774 the work has been constantly increasing in value and importance. There are upwards of forty varieties of cane, many of which have been first tried in the gardens, and the same is true of very many others of the most valuable products. Probably two-thirds of the fruit, nuts, choice woods and economic or medicinal plants now grown in Jamaica were introduced from foreign count ies.

The light-houses are at Morant Point, Plumb Point and Folly Point. The first is the beacon that greets the stranger who approaches Jamaica by the usual course, and lights him on his way when he is depart-ing. The second is the light whose faithful ray shines like a star across the palisadoes and the harbor to Kingston.

# GOVERNMENT REVENUE AND THE CUSTOM HOUSE.

THE revenue of the government is from the following sources: admeasurers' fees, warehousing, wharfage, import and export duties, light-house dues, rum duties, licenses, pilot fees, stamps, property taxes, general internal taxes, which include market fees and petty taxes of various kinds.

We have not space to enumerate the articles which are dutiable, nor those which are on the free list. Broadly it may be stated that food supplies, with few exceptions; all live stock except asses; wood and lumber; clothing and furniture; spirits, wines, etc., are subject to import duty either specific or advalorum. But mills and machinery or parts of the same; tools and implements used in the trades or in agriculture; most raw material for manufacture, etc.; books and works of art, are upon the free list. Sugar, rum, coffee, logwood and other valuable woods are subject to a small export tariff.

The regulations governing the inspection of baggage at the custom-house are as follows:

"On the arrival of each steamer alongside the wharf, passengers' luggage is removed by the company's servants, to the baggage room or floor for examination, free of charge. The customs are not responsible for lost luggage except it be removed to the Queen's warehouse. If passengers have any tobacco, gold or silver plate, wine, spirits of any kind, or any articles other than for their own personal use, it is necessary that they should declare the fact previously to the examination of their luggage. If this be neglected and any of the above mentioned articles are found, visitors may be subjected to the inconvenience of a thorough examination and the possible detention of all their baggage. The importation of merchandise with baggage is strictly prohibited."

# EXCHANGE, COINS, ETC.

THE Bank of Nova Scotia will buy and sell exchange on United States and Canada at rates governed by New York rates for sterling exchange.

The following is a list of the gold and silver coins in use and recognized in Jamaica :

| GOLD. | | | | SILVER. | | | |
|---|---|---|---|---|---|---|---|
| | £ | s. | d. | | £ | s. | d. |
| Doubloons— | | | | English— | | | |
| Old Mexican average.. | 3 | 4 | 0 | Crown ............... | 0 | 5 | 0 |
| Columbian ..... ..... | 3 | 0 | 0 | Dollar ............. | 0 | 4 | 0 |
| Aliquot parts in pro- | | | | Half crown......... | 0 | 2 | 6 |
| tion. | | | | Florin............... | 0 | 2 | 0 |
| | | | | Shilling............. | 0 | 1 | 1 |
| Sovereigns— | | | | Sixpence............ | 0 | 0 | 6 |
| English and Australians | 1 | 0 | 0 | Fourpence .......... | 0 | 0 | 4 |
| Half-Sovereign ditto.. | 0 | 10 | 0 | Threepence......... . | 0 | 0 | 3 |
| | | | | Twopence........... | 0 | 0 | 2 |
| American— | | | | Penny-halfpenny ..... | 0 | 0 | 1½ |
| Double Eagle........ | 4 | 2 | 0 | Nickel Coins— | | | |
| Single ditto.......... | 2 | 1 | 0 | Penny............... | 0 | 0 | 1 |
| Half ditto.. ......... | 1 | 0 | 6 | Half-penny.......... | 0 | 0 | 0½ |
| Quarter ditto......... | 0 | 10 | 3 | Farthing ........... | 0 | 0 | 0¼ |
| Dollar ............. | 0 | 4 | 1 | | | | |

Accounts are kept in £ s. d.

Weights and measures are the same as those used in Great Britain.

American money passes everywhere in business. The paper currency of the island consists of the notes of the Colonial bank. There is a nickel currency of pence and halfpence.

# CONCLUSION.

THERE have been many things necessarily omitted from this little work. Matters that we had planned to discuss, data which it is difficult not to think quite as important as much that we have published. Our omissions have been unavoidable; indeed, to have fully exhausted our note books and sources of information would require not one such volume as this, but twenty.

We abandon our purpose of furnishing a chapter on folk lore with a sigh, and regret the shelving of much entertaining material on the still potent superstition of obiaism. Only briefly and unsatisfactorily have we touched the subject of education; its advance and power in Jamaica demand more space then we could give it. But if we begin the catalogue of our delinquencies, where will they end? We must leave them to our critics, hoping that they will find in this book, even in its incompleteness, the evidence of honest purpose and a measure of fulfilment.

# INDEX

www.ingramcontent.com/pod-product-compliance
Lightning Source LLC
Chambersburg PA
CBHW021520210326
41599CB00012B/1329